Alien Skies

Frédéric J. Pont

Alien Skies

Planetary Atmospheres from Earth to Exoplanets

 Springer

Frédéric J. Pont
University of Exeter
Exeter, UK

ISBN 978-1-4614-8553-7   ISBN 978-1-4614-8554-4 (eBook)
DOI 10.1007/978-1-4614-8554-4
Springer New York Heidelberg Dordrecht London

Library of Congress Control Number: 2013958232

Springer is part of Springer Science+Business Media (www.springer.com)

Acknowledgments

This book grew out of the Exoclimes 2010 conference in Exeter, and I am grateful to the invited speakers for having accepted our invitation, and indirectly for providing the scientific material for this book: Fred Taylor, Sushil Atreya, David Grinspoon, Peter Read, Heather Knutson, Joe Harrington, Jonathan Fortney, Adam Showman, Christen Menou, Linda Elkins-Tanton, Frank Selsis, François Forget, Ralph Lorenz, James Kasting, Tim Lenton and Peter Cox, as well as to Suzanne Aigrain and Isabelle Baraffe for helping to make the conference happen, and Nawal Husnoo, Eli Bressert, Rob Da Rosa, and Joanna Bulger, for their help in the recording and webcasting of the conference. This book has had a long gestation period, and many people have contributed. First and foremost, I would like to warmly thank the editor, Robin Rees, who transformed that vague plan of mine into a real book, and who trusted and encouraged me every step of the way. I also thank Tom Spicer and Harry Blom, who promoted the project at Springer.

I am very indebted to Derek Christie and Tom Evans, who have devoted a lot of work to correct my sometimes erratic non-native English and improve the clarity of the explanations, and whose general comments were essential. I thank Jenny Patience for continuous support for this project, and Aude Alapini and Jo Barstow for their help with the home-made illustration.

I also thank the sources of inspiration for some of the drawings that I was not able to trace for inclusion in the image credits, including a splendid shop window installation in Zermatt and a comic book on dinosaurs with magnificent clouds.

All illustrations without credits in their captions are by me and the artist Jamie Symonds.

This project was financially supported by the Halliday Foundation and the Leverhulme Foundation, to which I express my deep gratitude.

Exeter, UK   Frédéric J. Pont
January 2014

Professor Frédéric Pont teaches planetary science at the University of Exeter, UK

Contents

Prologue

On the planet Isis, when the weather turns bad, a hail of red-hot glass droplets flies in the air at the speed of sound.

Isis is a gas giant planet orbiting an orange star in the constellation Vulpecula, some 60 light-years away from us. It is similar to Jupiter in size, but orbits very close to its sun, bringing its atmosphere to 1,000 degrees Celsius and higher. At these temperatures, rocks can vaporise and condense again as a rain of sand and glass.

The extreme temperatures also power very fast winds in the atmosphere of Isis[1], giant versions of the jet streams on Earth. Near the Equator they can be accelerated to velocities higher than the speed of sound. These supersonic currents produce shockwaves unlike anything on Earth, blasting the glass clouds in front of them.

There are no pictures of the planet Isis, and no probe has explored its atmosphere. What is known about it comes from indirect measurements from Earth. For instance, the presence of a haze of glass grains is inferred from measurements of the scattering of the light of the star by the atmosphere of the planet, observed in 2006 using the Hubble Space Telecope during the transit of the planet across the disc of the star. The type of scattering observed is characteristic of transparent drops, and a certain type of glass, amorphous silicates, is a very good candidate to explain the nature of these drops. The temperatures in the atmosphere were measured with the Spitzer Space Telescope by recording how the brightness of the planet in infrared light evolved along its orbit. The temperature varies between 1,220 degrees Celsius on the side of the planet facing the star and 700 degrees Celsius on the dark size. The speed of the wind on the planet was measured using the Very Large Telescope at the European Southern Observatory in northern Chile, by detecting tiny changes in the wavelength of a spectral line of carbon dioxide, superimposed by the atmosphere of the planet on the light emitted by the star as the planet passed in front of it.

This book is about the nascent science of planetary atmospheres beyond the confines of our Solar System. The science of extra-solar planets, or exoplanets, is very young. The study of their atmospheres is even younger.

Then again, as an imaginary pursuit, speculating about other worlds in the sky is

---

[1] The planet orbits the star HD 189733, hence it is called HD 189733b by astronomers. On the topic of planet naming, see note on page 138.

as old as civilisation itself. In the first century B.C., Roman author Lucretius mused about the infinite number of other worlds in the cosmos. Faraway planets became one of the recurring themes of science-fiction, but only at the end of the twentieth century were the first exoplanets finally detected around other stars.

The first confirmed planets outside the Solar System were found in 1992 around a pulsar, and the first exoplanet orbiting a solar-type star was announced in 1995. My own interest in exoplanet research took some time to develop. I was rather unmoved by the discovery of the first exoplanet. For one thing, it had always seemed quite obvious that planets were common around stars, and that finding one was long overdue. But the main reason was that the discoverer was just along the corridor from my office and was my PhD supervisor. Epoch-changing discoveries are not supposed to happen in the office next door, that is simply a matter of statistics and common sense, so I wondered whether the whole thing was true.

I was not the only one. There had been many false alarms in the hunt for exoplanets, dating back at least to the middle of the nineteenth century, and the scientific community had become cautious about any claims. In 1997, new results from the spectroscopy of exoplanet candidates – the technique used to detect exoplanets – seemed to cast doubt over the whole issue. Extra-solar planets are not detected directly; their presence is inferred from their effects on their host star. They tug on their star as they move along their orbit, and it is possible to detect the resulting wobble in the star by measuring the so-called Doppler shift of the lines in the spectrum of the light emitted by the star. However, unusual oscillations originating in the stars themselves could cause similar shifts, without there being the need for a planet.

But two years later, the detection of the transit of planet Osiris (HD 209458b) in front of its parent star clinched the case. Nothing could mimic the sharp signal of a planet passing across the disc of its star, a sudden drop in brightness followed about two hours

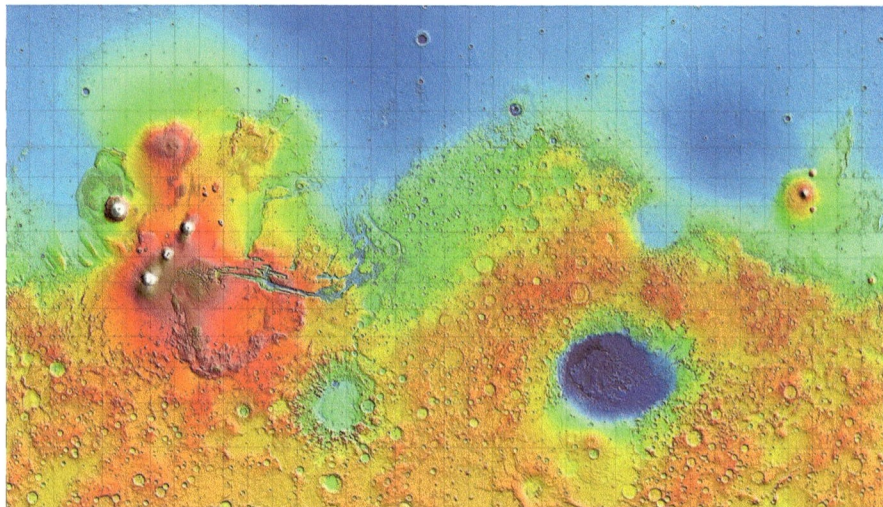

Fig. 0.1 Altitude map of Mars measured with the Mars Orbiter Laser Altimeter. Image credit: NASA

later by a sharp rise. This event was first caught with a small telescope at an American university, then measured with exquisite precision by the Hubble Space Telescope.

Still, with only mass, size and orbital period being known for these invisible planets, it seemed to me that a very long time would elapse before anything detailed would become known about them, about their appearance, atmosphere or weather, about anything that would allow us to "get personal" with any planets outside the Solar System, which is what I was really after. Luckily, I was completely wrong.

My own exoplanetary epiphany, strangely enough, came not from an exoplanet, but from Mars. In 2003, the Mars Orbiter Laser Altimeter (MOLA) measured the topography of Mars in exquisite detail, and the MOLA team produced the map of the red planet opposite.

This is when I belatedly realised that other planets were not abstract concepts in catalogues and scientific articles, but actual places. They had ravines, creeks, cliffs, promontories, gullies, slopes, small hills, wide plains and gloomy caves; as many interesting places as on Earth, perhaps even more. In their minute details and suggestive colours, the MOLA maps invited the armchair planetologist to endless adventures, scrambling down the ravines of Melas Chasma, navigating Kasei Valley, crossing the vast plains of Elyseum, crater-hopping in the northern reaches of Tempe Terra and sunbathing at the edge of Syrtis Major.

Maybe because the Apollo Moon landing occurred when I was only two years old, the black-and-white sequences with people bouncing in hermetic suits never gave me an intimate feeling of the Moon as a place. Also, because it does not have any atmosphere, the Moon is unlikely to become a real "place" for anybody in the foreseeable future. Mars is different. In a few centuries, a human could look out from the parapet of a terrace built with red rocks. A real person, not an intrepid astronaut but someone of a contemplative inclination, could be there on Mars and feel the cold wind on his or her skin, and love those stones with the kind of love that one can only feel for one's home planet.

I was left wondering what turns planets into places. What were the reasons that the Moon landings, planetary probes and science fiction literature had left me unenlightened. In spite of their technological veneer, the worlds of science fiction have an imaginary quality that make them more akin to fairy tales teeming with elves and dragons. As for the Moon landings, there is something about being locked in tin boxes and pressurised suits that cuts off the intimacy of the experience. It feels like exploring a deep-ocean trench in a pressurised vessel, when you would rather freely swim between corals. The dry, mineral and airless surface of the Moon is difficult to love.

In the end, what makes a place a place may have a lot to do with the presence of an atmosphere: the air, the breeze, the clouds. The stunning appeal of the Martian views that were beamed back to us by the probes and rovers lies in the sunsets, the drifting clouds, the morning frost. A place is a place if you can feel the wind and hear the sounds.

Of course, the atmosphere of Mars is extremely different from that of Earth. It is not breathable, and the pressure is so low that a pressurised suit is still required for human visitors. The atmosphere of our other neighbour, Venus, is even more alien, with a temperature hotter than molten lead and a pressure similar to the bottom of the sea. Given that our immediate neighbours are so different from us, and from each other, a question springs to mind; how many types of atmosphere are there?

It is tempting to suspect an amazing variety of possibilities. Indeed Titan, one of Saturn's moons and endowed with a thick atmosphere, is completely different again, numbingly cold and shrouded in a thick fog of oily droplets. The variety of atmospheres on exoplanets could be mind-boggling.

It wasn't until 2004, long after finishing my PhD, that I started working on exoplanets. The years since then have seen an explosive growth in the field. Dozens, then hundreds, of exoplanets were found. But perhaps most excitingly, at least from my point of view, this period has seen the first steps being made in studying the atmospheres of exoplanets. Atoms, molecules and aerosols have been detected in the atmosphere of a small number of the most favourable exoplanets, and the direction of planet-circling winds has been measured.

Nonetheless, the data that we have for exoplanetary atmospheres is still very limited, and quite a lot of it is considered dubious. Much of the slack has been picked up by the climate modellers, who have taken the sparse data that we have and used these to train their computer simulations and flesh out a more detailed picture of the atmospheres we are studying. Climate modelling already has a distinguished tradition for Earth, and a growing number of practitioners are starting to apply their methods to exoplanets.

In 2009 I attended a talk by prominent climate scientist Ray Pierrehumbert in Oxford, where I learnt that this effort was already well under way in the United States. One of the things Ray had been doing was calculating the possible climate of an Earth-like planet tidally locked to its sun, i.e. one that is always presenting the same side to it, in the same way that the Moon is locked to the Earth. He found that, in some cases, the climate could settle in a "pool planet" configuration, with the entire ocean frozen over, except at one opening on the side facing the star.

Inspired by this talk, I teamed up with colleagues to organise an international conference in 2010 at the University of Exeter in England, where I have been based since 2008. The objective was to bring together specialists on the atmospheres of Earth, Venus, Mars, Titan and exoplanets. "Exoclimes 2010" succeeded in this objective and attracted participants from all over the world.

This book largely draws from the content of that conference. Starting with Earth, the conference extended to neighbouring Venus and Mars, then to Titan and the giant planets in the outer reaches of the Solar System, before moving to the exoplanets. Among these, we first considered those for which present knowledge is more solid, the so-called "hot Jupiters", then we discussed the more speculative "super-Earths" and Earth-like exoplanets, that are yet to be reached by precise observations.

In 2012, a second "Exoclimes" conference took place in Aspen, Colorado. The new field of comparative planetology outside the Solar System, which sounded slightly dreamy in 2010, was already coming of age, with many participants involved in research on both Solar System planets and exoplanets.

Let us now begin our tour of these planetary atmospheres. We will move from the familiar (Earth) through the well-explored (Venus, Mars and Titan) to the barely known (hot Jupiters) and the speculative (other exoplanets).

– Prologue

Note on appendices

Complementary information, such as the names of the researchers involved in the discoveries, the scientific articles, and possible further readings on the subject of individual chapters, can be found at the end of the book.

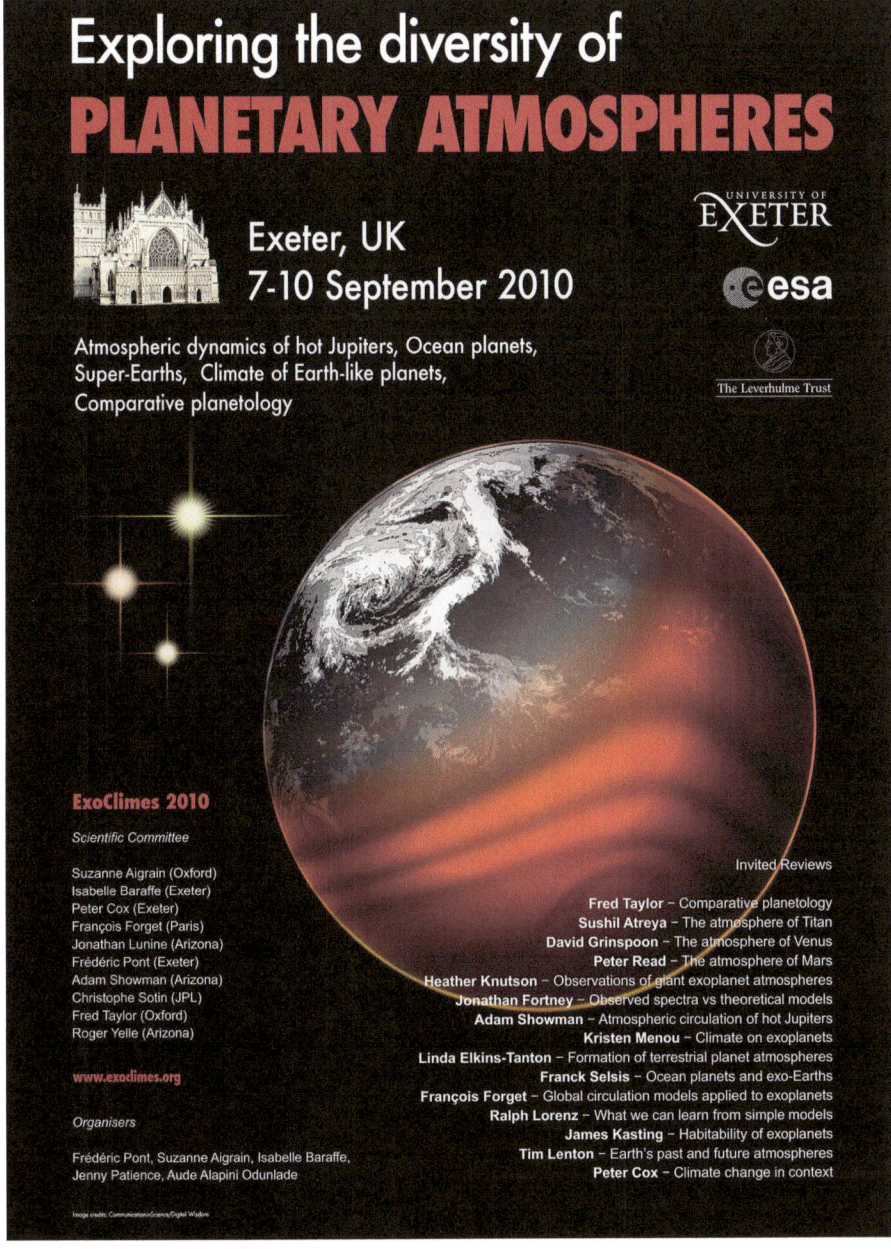

Fig. 0.2 Poster of the Exoclimes 2010 conference.

Chapter 1
The Home Planet

Earth's atmosphere: a new look at an old friend

We think we know quite a lot about the Earth's atmosphere, because that is where we happen to live. We breathe the air, we feel the breeze and the gales, marvel at the clouds swirling overhead. Rain and snow pace our seasons, and twenty-first century adults are well aware of the invisible layer of ozone protecting us from solar ultraviolet light, the carbon dioxide controlling the climate through the greenhouse effect, and the cloud-free and stable stratosphere above 12,000 metres (40,000 feet) where aeroplanes find refuge from turbulence. Although humans descended from oceanic creatures that ventured onto land – and consist mainly of water – the atmosphere is our new home.

Let's stand back a bit and take another look at our atmosphere, and ask ourselves what we really know about it. Maybe, as when an old friend does something out of character, we will realise that we didn't know as much as we thought, and that we have been mistaking familiarity for understanding.

For example, the gases in our atmosphere are invisible, odourless and tasteless, so as young children, we don't even think of air as a thing, we see no real distinction between air and empty space. At age seven, my son was still baffled by the way water rises inside an upside-down glass, slowly raised from a full bathtub. He laughed at the idea that the air in the atmosphere could push the bathwater downwards so strongly that it would rise in the glass. Much of our intuitive understanding of the atmosphere comes from the fact that when water condenses into droplets and ice crystals to form clouds, these clouds scatter light and become visible as white patches. Clouds, mists, storms, hurricanes – this is the atmosphere in action – are only visible because of the presence of water droplets. However, water is a minor component of the Earth's atmosphere, comprising less than one percent overall, and most of that is also invisible. The high-altitude currents that redistribute the heat from the tropics up to the poles, the ozone layer and stratospheric heating, the greenhouse effect, the carbon cycle, the oxygen disequilibrium, and many other atmospheric phenomena, all are invisible to our eyes.

Let us take another look at our old friend, the Earth's atmosphere, this time with the tools of science. With those we can see the invisible, and we might be surprised.

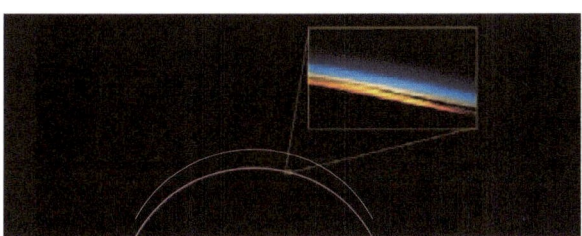

Fig. 1.1 Earth viewed from space. Half of the atmosphere is contained in a thin layer of 5.5 kilometres (3.5 miles), less than one thousandth of the radius of the planet. Image credit: NASA

The first surprise is how thin the atmosphere is on our planet. Looking at the dark side of Earth from space, the atmosphere is just the slightest of blurs near its edges, and even that is something of an illusion, because gases take up a lot of room compared to their weight. Were we to cool the Earth enough to liquify the air (which would require a temperature of about –200 degrees Celsius), the whole atmosphere would amount to a liquid layer only 14 metres (46 feet) thick, hundreds of times less than the depth of the oceans. Curled up in a ball in space, the whole matter in the Earth's atmosphere would measure about 80 kilometres (50 miles) across.

Fig. 1.2 The atmosphere frozen out. The pool of liquid nitrogen is 14 metres thick, the height of a five-storey building.

Why is the Earth's atmosphere so thin? Is this typical for such a planet? Looking at other planets in the Solar System, it seems that there is no such thing as a "typical" size for an atmosphere. The atmosphere on Venus is about one hundred times denser than here, and the one on Mars is one hundred times thinner. Some planets have no atmosphere at all. At the other extreme, the giant planets have an atmosphere so deep that they have no solid surface.

We will find as we journey through this book that the diversity of atmospheres has a lot to do with the chaotic nature of the birth of planets. In fact, one question is why the Earth should have an atmosphere at all. Planets made of rocks and metals like the Earth form through cataclysmic collisions of smaller bodies, and the heat from these collisions is more than sufficient to vaporise any nascent atmosphere back into interplanetary space within the first few million years of their turbulent birth.

Geoscientists now think that the Earth's atmosphere is not its initial wrapping of gas, but was built up later on from the exhausts of volcanic eruptions and lava flows. The gases in our atmosphere were trapped in rocks deep inside the mantle of the planet, and gradually made their way out over time. The atmosphere of our planet is but a thin layer of volcanic fumes exuded by the rocks of its interior in earlier eras.

Fortunately for us this is not the whole story. The smoke from volcanoes contains mostly carbon dioxide and acid sulphur compounds, a very different mix from the present composition of the atmosphere.

Which leads us to the second surprise about the Earth's atmosphere, its composition.

Our atmosphere contains one of the most powerful explosives. Air is made up of four parts nitrogen to one part oxygen, with a little water, argon[2] and carbon dioxide, not to mention a whiff of methane.

Fig. 1.3 The composition of Earth's atmosphere: a mixture of nitrogen and oxygen with a sprinkling of water (0.5%), carbon dioxide (0.03%) and argon (1%).

Nitrogen, water, argon and carbon dioxide are the kinds of gases one would expect to find in a planetary atmosphere. They are simple molecules made of some of the most common elements in the Universe. But the sheer abundance of O_2 would be puzzling to an alien visitor. This is because O_2 is a highly reactive molecule, and almost anything left in contact with it for a long time will end up oxidised, or in other words, it will burn or rust[3]. Oxygen is removed from the air in the process. Since there is more than

---

[2] Yes, argon is the third most common element in the Earth's atmosphere. We hear very little about it because it is chemically inactive (a "noble gas").

[3] As in the apocryphal story about the Eskimos having hundreds of words for snow, we have many words to describe the process of oxidation, depending on the speed at which it proceeds, but the basic phenomenon

enough stuff to oxidise on Earth to soak up all the oxygen in the atmosphere, the presence of such a huge reservoir of free oxygen in the Earth's atmosphere would come as a shock to the innocent visitor. Finding O_2 in such high concentration in a planet's atmosphere is like opening the fridge and finding that most of the yoghurts are full of dynamite.

Fig. 1.4, 1.5 The slow oxidation of the Titanic, and the rapid oxidation of the Hindenburg. The two most glamorous catastrophes of the twentieth century ended in the oxidation of large amounts of material.

Indeed, it is the highly combustible nature of oxygen that makes the existence of hyper-active creatures like us on this planet possible. Over billions of years of evolution, our bodies have become efficient factories for taking in un-oxidized matter from our surrounding environment in the form of food, and using the air's oxygen to burn it. This process will power your body to run a marathon and carry your car around the world ten times over. Other sources of energy, such as photosynthesis, produce much less power than oxygen-burning. This is why plants grow slowly, and why there are no green photosynthetic animals running around.

The flip side of the same coin is the low abundance of CO_2 in our atmosphere (0.3 percent). This molecule is the signature product of volcanic activity, and tends to stay around for a long time. We can see it in the atmospheres of Venus and Mars, where carbon dioxide is the main gas. The fact that Earth's atmosphere is far richer in unstable O_2 compared to the relatively inert CO_2 is obviously saying something special has been happening to this planet.

Another surprising feature of our atmosphere is the way it moves. The dominant motion in the Earth's atmosphere is a colossal roll between the Equator and the tropics. Air rises from the ground and oceans near the Equator, dropping its humidity in the process in torrential downpours over the rain forests of the Amazon, the Congo basin and South-East Asia. Travelling sideways towards the tropics, the air then rushes downwards and back to the Equator in the guise of burning dry winds over the deserts of the Sahara, Australia, Persia and the Kalahari.

---

is the same: things burn, explode, rot or rust. On this planet, if you are not oxidised already, your days are numbered. This is valid for us as well, our body uses a lot of anti-oxidant to slow down this process, including vitamin C. That is also why petrol and natural gas are buried deep underground, and have formed only rarely during the eons of life's passage on earth. The formation of these fragile carbon-rich compounds required oxygen-starved conditions. We tend to think of petrol, coal and natural gas as fuel; but in fact, the real fuel is the oxygen in the atmosphere, fossil fuels only provide the sacrificial lamb for this starved molecule to feed on. Without oxygen, petrol is useless.

The rising of the atmosphere ends rather abruptly at about twelve kilometres, above which the air is stable and does not rise or fall. Although it does not move vertically, air does, however, flow around the Earth at high speeds, in large snakey currents that circle the whole planet. These currents are the "jet streams" that Bertrand Piccard and Brian Jones used when they made the first circumnavigation of the Earth aboard a hot air balloon in 1999.

Nearer to the poles, in the temperate regions where a large fraction of humankind lives, air below ten kilometres (six miles) altitude oscillates in elegant swirls, a few thousand kilometres in size, producing the "weather systems" that show up on the evening news.

Fig. 1.6 Profile of the Earth's atmosphere. The vertical scale shows an equal interval each time the pressure doubles. At sea level the pressure is one bar by definition. At around 12,000 metres (40,000 feet), convection stops and the air becomes stable (stratosphere). Ten metres (30 feet) below water, the pressure is two bars. At the bottom of the oceans it reaches one thousand bars. The highest human settlements are near the half-bar mark.

The role of pressure

At a given altitude in the atmosphere, pressure is simply proportional to the amount of material sitting overhead. On Earth at sea level, the average pressure is one bar (the standard unit of atmospheric pressure), which corresponds to one atmosphere's worth of air above one's head, or ten tonnes of air per square metre of surface. It also corresponds to the pressure of 981 millimetres (38 inches) of mercury, or a column of around 10 metres (30 feet) of water.

Owing to the way gases react to pressure, most of the Earth's atmosphere is concentrated near sea level. Although very thin layers of air extend out to altitudes of 100 kilometres or more, half the mass of the Earth's atmosphere lies below an altitude of 5.5 kilometres (3.5 miles). As tourists try to catch their breath in the narrow streets of the Peruvian city of La Rinconada, fifteen thousand feet up on the Andean Altiplano, it is worth pondering that they are already halfway up into space.

Fig. 1.7 La Rinconada in Peru stands 5,000 metres (16,700 feet) above sea level, and the pressure there is about half a bar. In other words, half the Earth's atmosphere is below the town, and the shallow-breathing inhabitants are halfway into outer space. The thirty thousand inhabitants live from gold mining. Permission: Hildegard Willer

At an altitude of 50 kilometres (30 miles) the pressure is a thousand times lower than at sea level, at one millibar, meaning that 99.9 percent of the atmosphere's matter is below this level. At 100 kilometres (60 miles), the pressure is down to one millionth of a bar. The 100 kilometre mark is taken as the beginning of "space" for the purpose of orbital tourism, although Earth's atmosphere does not have a sharp boundary, but very gently dissolves into the extremely small densities of the solar wind and interplanetary medium that permeate the whole Solar System.

Our intuition doesn't give us a good understanding of how gases behave in terms of pressure, because air is invisible and our body can adapt to changes in pressure. Thinking about sea diving is a good way to understand pressure in planetary atmospheres. Water is a nearly incompressible liquid, and under water the pressure is simply proportional to the depth. Ten metres (33 feet) below the surface, the pressure has increased by one bar, at 20 metres (66 feet), two bars, etc... The pressure grows with the weight of water overhead, just like a pile of bricks is twice as heavy if it is twice as high. The world record for scuba diving is 330 metres (1,000 feet), where the pressure is a whopping 34 bars. Pressure does not change the shape of divers (nor fish or whales), because animal bodies are mainly made of incompressible fluids. However if a diver carried a balloon full of air to such depths, it would be compressed to 1/34th of its original size. At the bottom of the deepest ocean trenches the pressure exceeds 1,000 bars.

Liquids and gases react very differently to changes in pressure. Because liquids do not shrink much or heat up when pressure is applied, conditions at the bottom of the oceans are not very different to those near the surface; the pressure is intense, but the temperature remains cool. Air does not behave like this. When pressure increases gases shrink to occupy less volume. As a result, the same mass of atmosphere will occupy much less space near the ground than at higher altitudes, one thousand times less space if we compare the one bar level (sea level) with the one millibar level (50 kilometres up). Mathematically, this implies that the pressure decreases exponentially with altitude in a planetary atmosphere, which means that it drops by the same factor at fixed intervals.

To visualise this phenomenon, imagine air parcelled into separate bags. The bags at the bottom are crushed by the weight of the bags above and therefore they shrink. As

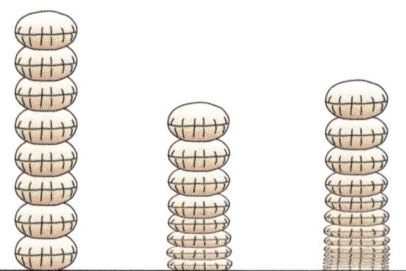

Fig. 1.8 When parcels of air are stacked on a planet, they do not pile up like bricks (left), but the pressure on the lower parcels makes them shrink (middle). An atmosphere with twice the air (right) is not twice as thick, because the lower layers are compressed by the weight of all the air above. Mathematically the pressure decreases exponentially with height.

more bags are added the atmosphere does not become proportionally thicker, because the weight of the new bags makes all the others shrink.

In practice, this behaviour cannot be illustrated in a laboratory or lecture hall, since gases are very light, and the bags need to be kilometre-sized for the pressure effect to become visible.

The highest air pressure point on the surface of the Earth at present is the coast of the Dead Sea, 360 metres (1,200 feet) below sea level. The air pressure there is a few percent higher than at sea level – which is not so remarkable since at a given place the pressure can fluctuate by such an amount due to changes in the weather.

If the Colorado Grand Canyon, for instance, had its rim at sea level, then at the bottom of the massive trench, two miles below, the pressure would be 50 percent higher than at sea level. The temperature would be around 15 degrees higher than at the edge of the Canyon because of the pressure. On the plus side, aeroplanes would find it very easy to glide on the thick air, and parachuting with an umbrella like Mary Poppins might even be attempted.

The role of temperature

Mountain-climbers know that pressure is one of the two key parameters in an atmosphere, the other being temperature. Pressure tells how much push comes from the outside,

Fig. 1.9 Gold is also the reason for men to visit the other pressure extreme, the bottom of the East Rand mine in South Africa, with galleries extending more than 2,000 metres under sea level. The pressure there is around 30 percent higher than at sea level (about the pressure at the bottom of an Olympic swimming pool), and the temperature at least 15 degrees Celsius higher than near the surface. Image credit: National Geoscience Database of Iran

temperature is a measure of how quickly the atoms and molecules move around and how hard they hit each other. Temperature affects our well-being by slowing down or accelerating the chemical reactions in our tissues, and pressure upsets the balance between the inside of our body and the outside.

The mean temperature on Earth at sea level is about 15 degrees Celsius. The mean temperature generally decreases with altitude in a predictable way, set by the laws of gases. When the pressure decreases a normal gas cools, and the temperature and pressure changes are proportional. These gas laws are fundamental to many familiar applications, like the steam engine (heating water vapour to increase the pressure in the engine) and the refrigerator (lowering pressure in a gas to cool it). The temperature drops by one degree for every 150 metres (500 feet) of altitude or so. People who come from mountainous countries are familiar with this temperature profile.

But temperature does not always follow this one-degree-per-150-metre law, because air can move up and down to balance the pressure and temperature. This is *convection*, the rise of air heated near the surface. As the Earth is heated by sunlight, the hotter air near the ground becomes less dense and rises upwards, lifting up cumulus clouds, birds and paragliders.

Stratosphere and ozone layer

About ten kilometres (six miles) above ground level, something changes; the air from the ground no longer rises, because it meets a cap of warmer air. At that height, air temperature is around –60 degrees Celsius, but even higher up, for instance at 30 kilometres, the temperature has risen again to zero degrees Celsius. Because temperature is no longer falling with altitude, convection is suppressed. For this reason storm clouds rarely rise up above ten kilometres, the upper limit of the convective, turbulent part of our atmosphere. Commercial air flights rise above this level to avoid turbulence. The region of the atmosphere above ten kilometres, where the air no longer rises, is called the stratosphere because it is "stratified" in stable layers, unlike the turbulent zone below where "weather" happens.

The rise in atmospheric temperature within the stratosphere is due to the ozone layer. Ozone is a form of the oxygen molecule, with three oxygen atoms (O_3) instead of two, that forms high in the atmosphere under the effect of ultraviolet light from the Sun. Ozone in the stratosphere blocks the most energetic ultraviolet light from the Sun and prevents it from burning us and sterilising the surface of the Earth. We know this because certain industrial gases have attacked the ozone layer in recent decades, thus allowing harmful ultraviolet rays to reach the surface in some regions. A less well-known consequence is that in the process of blocking out UV light, the ozone acquires the heat that might have burnt our skin, and keeps it in the stratosphere; and this is why the stratosphere is warm. The ozone layer keeps the lid on the Earth's weather by reversing the decrease of temperature with altitude above ten kilometres. Without it, storms could rise much higher up in the atmosphere towards space, a magnificent sight no doubt, but a dangerous situation for creatures living on the surface.

Fig. 1.10 The stratospheric lid in action. A volcanic plume shoots through the cloud layer in Iceland. The volcanic gases and ashes are much hotter than the air near the ground, and even as they cool with the dropping pressure, they remain hotter and keep rising. But when the volcanic gases reach the stratosphere heated by ozone, their rise stops and they start spreading out with the winds. Image credit: NASA, ISS

Earth's double glazing

The ozone layer that blocks high-energy, ultraviolet radiation and confines weather to the lower layers is one of two invisible "glass panes" that equip the Earth's atmosphere.

The second invisible glass pane is made up of greenhouse gases lying closer to the ground, chiefly water vapour and carbon dioxide, which prevent infrared radiation (heat) from escaping back into space.

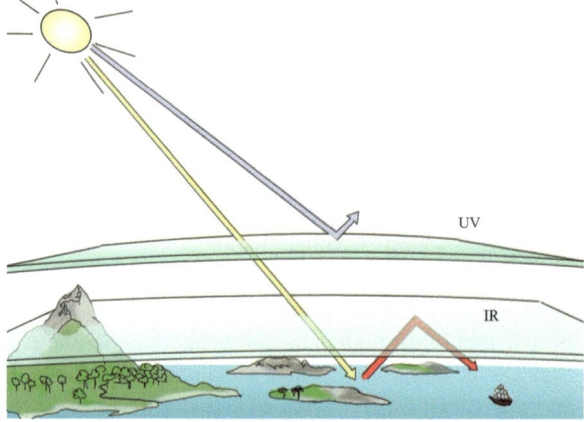

Fig. 1.11 The double-glazing of the Earth's atmosphere. The ozone layer blocks the incoming UV radiation, protecting life, and heating the stratosphere. The greenhouse gases keep the infrared radiation near the ground, maintaining the global temperature above freezing.

As a result, our atmosphere performs the trick of being almost perfectly transparent to visible light and opaque to harmful ultraviolet rays from the Sun, while being able to maintain the warming infrared rays close to the surface.

With a "sunblock" screen to keep harmful ultraviolet light out, an "insulation" layer[4] to keep heat in, and transparent air to make abundant sunlight available for photosynthesis, the plants covering the Earth couldn't have wished for a better atmosphere if they had designed the atmosphere themselves.

Of course, this is no accident, and in a very real sense, plants *did* engineer our planet's atmosphere.

Atmospheric circulation

The reaction of air to changes in temperature and pressure is the key to the way air moves around the globe to create weather and climate. The radiation of the Sun heats the air near the surface, most strongly in equatorial regions where the Sun is directly overhead. Because pressure increases with temperature in a gas, the hot air expands and it becomes less dense, which causes it to float upwards (this is buoyancy, or Archimede's principle: something less dense placed into a fluid will be pushed upwards). As it hits the ten-kilometre limit, the air moves sideways towards the cooler regions of the Earth. Then it hits another limit, set against polewards motion by the rotation of the Earth (more on this below). By now it has cooled down a bit, it becomes denser again and sinks downward. The loop is completed by the low-altitude trade winds flowing from the tropics and the Equator, which bring the air back to where it started.

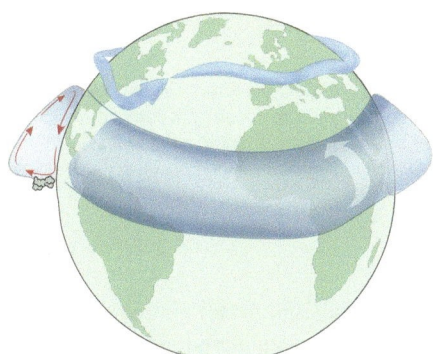

Fig. 1.12 Global atmospheric circulation on Earth: the Equator-to-pole churning, and the mid-latitude swirls.

The large churning motion of our atmosphere from Equator to tropics is the dominant movement in our atmosphere. However, it does not encompass the whole planet, because of a key physical fact on a spherical planet – the *Coriolis effect*.

The Earth spins quite rapidly on itself, with 365.25 turns on itself (days) during the time that it takes to complete one revolution around the Sun (year). Given the Earth's circumference – 25,000 miles – its one-round-per-day rhythm implies a speed of more than 1,000 miles per hour at the surface of the planet, on the Equator. Thanks to our puny size, we remain blissfully unaware of this breakneck speed. Only the sight of the Sun, Moon and planets coursing across the sky are there to remind us of what is going on.

---

[4] Every metaphor has its limits. Both ozone and carbon dioxide extend over a large proportion of the atmosphere, not the thin location suggested by terms like "layer" or "pane". Ozone is spread out between 20 kilometres and 80 kilometres, carbon dioxide is mixed throughout the atmosphere, its main effect concentrated on the bottom 10 kilometres.

This speed of rotation is much faster than the speed of the winds in the atmosphere, which means that, by the time it takes a stream of air to cross the Earth (usually a week or more), it will have turned around many times with the whole planet. This makes the rotation of the planet an essential factor in shaping weather and climate.

Since planets are spheres rather than cylinders, as one gets nearer to the pole, this velocity decreases. One step away from the North Pole, our rotation velocity would only be a few centimetres per hour. Like the Little Prince moving his chair on his tiny planet to enjoy the sunset many times, we could easily stroll around the pole to remain constantly at the same local time.

This drop in velocity from 1,000 miles per hour to zero when moving polewards is the cause of the Coriolis effect. Atmospheric or oceanic currents are liable to travel over distances comparable to the size of the whole Earth. When they try to move from the Equator towards either of the poles, they are leaving areas which are rotating eastwards at more than 1,000 miles per hour and edging towards areas which are rotating at much lower speeds. Unable to keep straight, they violently swerve eastwards, because of all the excess rotation velocity they are bringing with them.

Fig. 1.13a,b The archer is aiming at the apple on a rotating platform. His arrow will miss its target completely. Carried by the sideways motion of the platform, it will follow a straight line that will look like a bent arc on a "map" of the platform constructed by a projection around the rotation axis, the way world maps are drawn. By the same effect, polewards motion in a rotating planet is diverted eastwards.

At mid-latitude (the latitudes of Europe and the Continental United States), the Coriolis effect starts to dominate, preventing the warm equatorial air from reaching the polar regions. The mid-latitudes are dominated by swirling, unstable weather patterns, a permanent failed attempt to balance the temperature and pressure of the hot sub-tropical and cold polar regions of the planet, thwarted by the Coriolis effect. The Equator-to-tropic roll is stable, and as a result rain on the Amazon is as predictable as droughts in the Sahara; while the trade winds can be relied on by sailors year after year. But the mid-latitude cyclones and high-pressure systems come and go in a matter of days – the time it takes for the Coriolis effect to bend their paths out of shape and send the air masses in the wrong direction – causing the beautifully unpredictable weather that some of us love and cherish.

The Coriolis effect explains the typical duration and size of weather systems at mid-latitudes. It takes a few days for a mass of air to complete an arc around the width of the mid-latitude regions, which are 2,000 to 3,000 kilometres (1,000–2,000 miles) across. With a typical wind speed of a few dozen miles per hour, this is about the time it takes to cover the same distance by car.

Nearer to the poles, the Coriolis bend is so strong that it becomes very difficult for the air to move polewards. Air masses swirl around the polar axis, forming polar vortices. This isolates the poles from the warmer regions and allows the temperatures there to reach very low values. On Earth, the polar vortex is especially efficient near the South Pole, since it is reinforced by the presence of a large continent with a thick ice sheet, and an uninterrupted passage for the ocean all around the continent that can also isolate it from warmer sea currents. In the North, the shape of the continents favours an occasional incursion of warmth into the polar regions, either by air or by sea.

Water cycle

Along the looping Equator-to-tropic circulation we encounter another all-important phenomenon in our atmosphere – the water cycle.

Water circulates between sea, air and land as vapour, liquid or solid. Evaporation above warm oceans fills the sky with clouds, which rain down to form rivers or fall as snow on icecaps. This cycle constantly replenishes the atmosphere with water vapour, which coalesces into raindrops or ice crystals. The main engine of the water cycle is the large loop pictured in Fig. 1.14, combined with the fundamental physical principle that, as the temperature drops, air is able to take up less water.

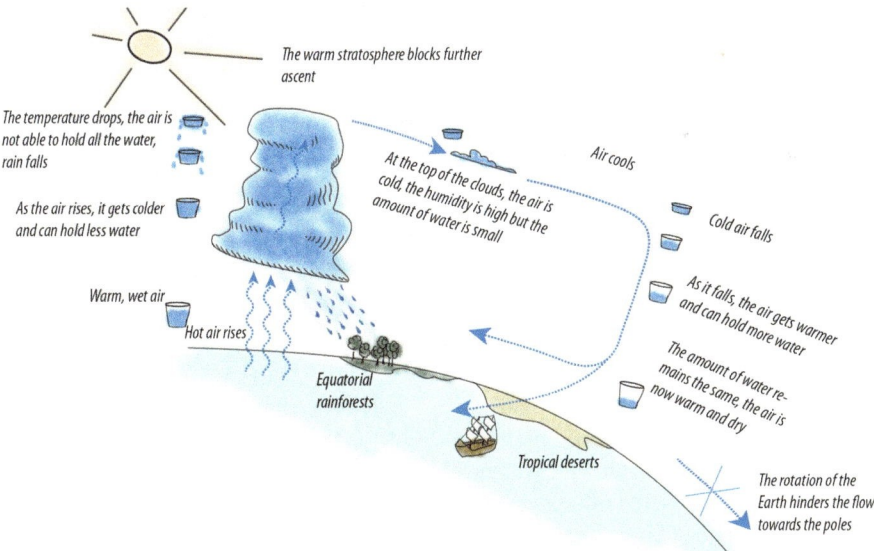

Fig. 1.14 The Equator-to-pole circulation of air on Earth and the role of water.

Indeed, water vapour can mix with air, but only up to a certain concentration, above which it starts to precipitate into droplets and form clouds. The amount of water vapour that air can hold depends very strongly on temperature; warm air can contain several percent water, cold air virtually none. For this reason, and somewhat counter-intuitively, clouds do not form when the quantity of water vapour in the air increases, but when air which is already humid becomes colder. This happens when a mass of warm, humid air encounters cold air; or when humid air is forced to rise upwards into colder regions.

Astronomer's chemistry

Nitrogen, ozone, greenhouse gases, water... the gases playing key roles in the Earth's atmosphere seem to keep cropping up without rhyme or reason. Fortunately, this is not the case, and we have now encountered almost all of the important molecules in planetary atmospheres, not only on Earth, but even in Solar System planets and extra-solar planets in general.

The fact is that, although there are 92 elements in nature and an incredibly vast array of molecules that these elements can form, the processes in stars that synthesise these elements vastly favour some, and make others extremely rare. As a result the "astronomer's periodic table" is very lopsided, and dominated by only a handful of elements. After hydrogen and helium, which were the only two elements synthesised in abundance by the Big Bang 13.7 billion years ago (and still vastly outnumber all other elements put together), the other common elements are carbon and oxygen (together accounting for most of the rest), followed by nitrogen, iron, magnesium and neon.

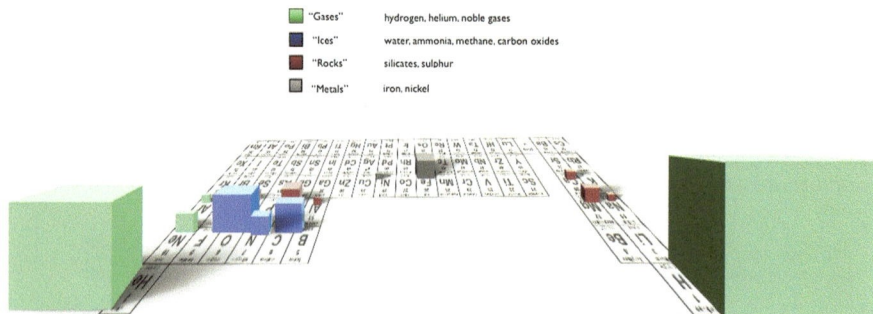

Fig. 1.15 The astronomer's periodic table of elements. The size of the boxes is proportional to the abundances (by weight) typical of the interstellar matter from which stars and planets are formed.

The role of the elements in planetary atmospheres depends on the molecules they can form. Hydrogen and helium are too light for the Earth's gravity to keep in the atmosphere; whatever amount was present after the formation of our planet drifted into space within its first billion years of life. Oxygen exists as water (H_2O), molecular oxygen (O_2, plus a bit of ozone O_3) and carbon dioxide (CO_2) while methane (CH_4), combines carbon with some of the hydrogen before it escapes into space. Nitrogen forms molecular nitrogen (N_2) and ammonia (NH_3). Neon, like argon, is a so-called "noble gas" because it considers that binding with other atoms is beneath its dignity.

Iron, magnesium, silicon and sulphur and the other heavier elements tend to form solids rather than gases, at least in the range of temperatures at the Earth's surface.

We are then left with only six key molecules for Earth-like planetary atmospheres: one rather passive compound (nitrogen), four greenhouse gases (water, carbon dioxide, methane, ammonia), as well as one very reactive gas (oxygen). As we will see in later chapters, these gases also dominate the atmospheres of other solid planets in the Solar System, like Venus, Mars and Titan.

Carbon-rock cycle

Apart from the very visible water cycle, the Earth's atmosphere hosts another huge cycle, much slower and more discreet than the water cycle – the carbon-rock cycle.

The interior of the Earth is made mostly of rocks in the mantle, and iron in the core. Rocks are compounds of silicon, oxygen and metals such as magnesium and aluminium. Looking again at the astronomer's periodic table of the elements, the global composition of the Earth now seems to make good sense. Helium and hydrogen are very light, which is why they have mostly escaped into space. The bulk of the planet is made up of condensed compounds of iron, silicon and other heavy elements, and there is oxygen and nitrogen in the atmosphere. The odd one out is carbon. Carbon most readily combines with oxygen to form carbon dioxide, CO_2, which is a gas at normal Earth temperatures. Shouldn't there be a lot more carbon dioxide in the air?

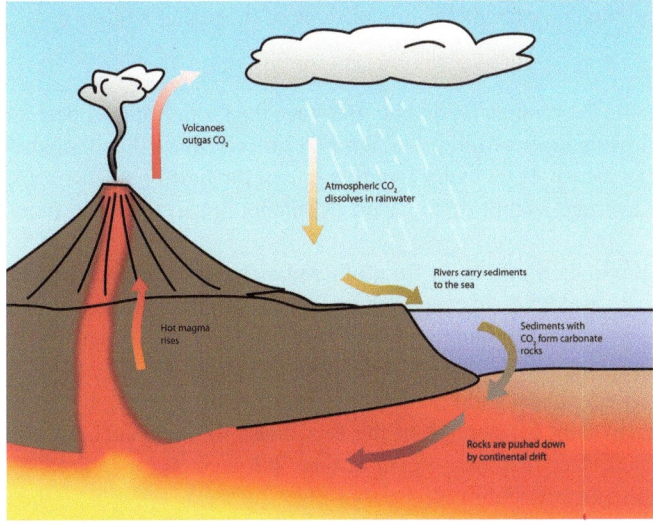

Fig. 1.16 The rock-carbon cycle.

Volcanoes outgas CO_2

Atmospheric CO_2 dissolves in rainwater

Rivers carry sediments to the sea

Hot magma rises

Sediments with CO_2 form carbonate rocks

Rocks are pushed down by continental drift

Carbon dioxide is constantly spewed out into the atmosphere by volcanoes, and over the 4.6 billions of years of Earth's life, there have been enough volcanic activities to fill the whole atmosphere with CO_2. However, our atmosphere contains less than one part per thousand of it, because carbon dioxide combines over time with other atoms to form sediment rock in the presence of water (the White Cliffs of Dover for instance are mostly made of calcium carbonate, $CaCO_3$). The Earth's oceans and rains serve as a go-between to transport the carbon back into the ground. The loop is closed by continental plates slowly grinding and pushing underneath each other. Plate tectonics bring sediments deep down into the Earth's magma, where the carbonates boil and free the CO_2 gas which makes its way up to the surface through volcanoes.

CO_2 is crucial to life on Earth, because its contribution to the greenhouse effect keeps the temperature relatively warm. Of course, if one species starts digging up carbon-rich compounds, burning them and releasing the resulting gases into the atmosphere, the greenhouse effect might get out of hand, but what are the odds of that?

The atmosphere as a machine

The Earth's atmosphere is, in a literal sense, a heat engine. In the same way that a steam engine uses moving vapour to transport heat from a hot place to a cold place and produce work, the atmosphere uses the temperature contrasts caused by the light of the Sun to move the whole atmosphere around in its ceaseless ballet; to power the winds, the waves and the storms.

The atmosphere is also a chemical reactor, baking CO_2 into rocks, and extracting oxygen from volcanic fumes, with more than a little help from its more liquid associates – the seas and living creatures.

Altogether, the atmosphere is a fantastically intricate and beautifully balanced machine, combining the interface between the cold, rapidly changing physics of interplanetary space, and the hot, extremely slow physics of the planet's interior.

The mighty ocean

But there is an elephant in the room, a liquid one. Between the Earth's atmosphere and the planetary rocks, flows a layer of fluid two hundred times heavier than the whole atmosphere – the global Ocean. Over every square metre of the Earth's surface sit ten tonnes of air, but the average weight of water on the planet is two thousand tonnes per square metre.

This really is what sets the Earth apart from other planets in the Solar System. If the Earth was warmer, the water would evaporate into the atmosphere. If it was cooler, the seas would freeze over and become part of the bulk of the planet. Thanks to the "just-right" temperature, water on Earth has settled into that particular state half-way between a gassy atmosphere and a solid planet – a liquid layer.

The oceans represent only 0.1 percent of the total mass of our planet, but they cover two-thirds of the surface to a mean depth of three kilometres (two miles). The oceans are key players in the circulation, chemistry and climate of the Earth's atmosphere. Sea currents transport about the same amount of heat from the warm equatorial regions to the cold poles as do the atmospheric winds and air currents. The ocean's waters mediate both the water cycle itself and the rock-carbon cycle. They control the climate by providing huge thermal inertia for islands and coastal regions, while leaving the centre of continents vulnerable to temperature extremes.

Sea currents are shaped by the position of the continents. Ocean currents link up to form a *global conveyor belt* configuration that snakes around the whole world.

Because oceans can store so much heat, these currents also have a profound effect on the circulation of the atmosphere. The Gulf Stream keeps northern Europe much warmer than corresponding latitudes in America and Asia, while the southern circumpolar currents confine Antarctica to a distinctly cold pocket.

It is remarkable that water is a liquid at room temperature, while other substances with similar molecular weights such as methane, carbon dioxide or nitrogen are gases

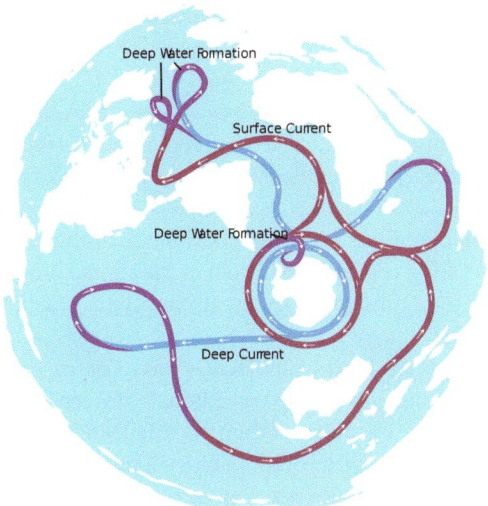

Fig. 1.17 The main ocean currents on Earth. Their features are dictated by the direction of the dominant winds and the shape of the continents. Image credit: AVSA

and only become liquid at far lower temperatures. As a general rule, lighter compounds condense last, with hydrogen and helium as the most extreme examples (helium freezes at –269 degrees Celsius, a mere four degrees above absolute zero, hydrogen at –252 degrees Celsius). Carbon dioxide (22 times heavier than hydrogen) becomes liquid at –57 degrees Celsius, nitrogen (14 times heavier) at –195 degrees Celsius, and methane (eight times heavier) at –161 degrees Celsius. Water (nine times heavier than hydrogen) is the odd one out.

Methane   Carbon dioxide   Water

Fig. 1.18 The shape of the methane, carbon dioxide and water molecules. Water is bent with electron-hungry oxygen on one side and electron-donor hydrogen on the other, making it much more inclined to bond with other molecules.

This is ultimately due to a small quirk of quantum physics. The shape of methane, nitrogen and CO_2 molecules is symmetrical, so identical atoms tend to face each other when two similar molecules meet. This is not a very powerful link, as atoms prefer bonding with atoms of opposite attitudes towards electrons – electron-hungry with electron donor. The water molecule, however, does not feature the two hydrogens on opposite sides of the oxygen atom, but with a kink of 105 degrees. This means that a water molecule can present its "back" to another, placing the hydrogen of one molecule near the oxygen of another. This is quite a strong link, which can keep molecules together despite the agitation of high temperatures.

The same little quirk is what makes water an excellent solvent, and thus an ideal liquid for life. Water molecules tend to act as little magnets that can take apart other molecules, such as salts (common salt, $NaCl$, readily dissolves in water into Na^+ and Cl^- ions). Living cells rely on this property of water to function, but that is another story... which brings us to the last great actor in the story of Earth's atmosphere – life.

The role of life

Life has many effects on the balance of the atmosphere. Vegetation makes the continents darker and allows more light from the Sun to be absorbed. Around 10 percent of all the water evaporated into the atmosphere comes from the respiration of plants, mainly in the equatorial rain forests.

But the main effect is the impact, over time, on the composition of the atmosphere itself. Plants (and algae) have been taking carbon dioxide out of the air and producing oxygen on a grand scale for such a long time that they are primarily responsible for the composition of today's atmosphere.

Like all life-forms on Earth, plants are built of molecules based on carbon, the most common of which are carbohydrates – made up of carbon, hydrogen and oxygen. Plants and algae build carbohydrates by taking carbon dioxide from the air, and water from the ground. They use solar energy through photosynthesis to break the oxygen-hydrogen bond inside the water molecule (not easy, as this is a strong bond), then the plants release the dangerously reactive free oxygen, and use the hydrogen to react with carbon dioxide and form carbohydrates. Animals live on the back of the plants in this respect. When consuming the plant's carbohydrates, they combine them with the free oxygen in the atmosphere, which is there thanks to the plants, to obtain energy (it may hurt our pride, but under terms such as "eating" and "breathing", we are merely undoing the patient chemical work of plants).

Over billions of years, plants have added a gigantic amount of free oxygen to the atmosphere and taken out a corresponding amount of carbon dioxide. The amount of oxygen in the atmosphere has been stable at around the 20 percent mark for hundreds of millions of years. The oxygen level is set not only by the amount that plants can produce, but also by how fast other processes can remove it from the atmosphere. These twin processes are linked through feedback loops; for instance, higher oxygen levels in the atmosphere would make forest fires much easier to start and propagate, which burns more plants into carbon dioxide.

Life also influences the composition of the atmosphere through the carbon-rock cycle. The total biomass on the Earth at present amounts to 500 billion tonnes, about half the total weight of CO_2 in the atmosphere. The carbohydrates in a plant are only sequestered from the atmosphere for the duration of its life. When the plant dies, it decays back into carbon dioxide. However, the return of CO_2 to the atmosphere can be prevented in certain cases. Firstly, if there is not enough oxygen available locally, microbes and animals cannot decompose the plant matter, which remains stored as carbohydrates. This is what happened in oil and gas deposits, where huge pools of plant matter have decayed in oxygen-starved conditions[5]. Secondly, many animals in the sea have resorted to using calcite shells as a protective device. To build them, they combine carbon dioxide with calcium ions in water to form calcite, a rock-hard compound that sinks to the bottom of the sea at the end of their life. Like many sediment formations throughout the world, the White Cliffs of Dover are made mainly of ancient organic matter, and they have sequestered a lot of carbon dioxide from the atmosphere.

Whereas life is entirely responsible for the oxygen in the atmosphere, it plays a secondary role regarding the amount of CO_2, which is mainly controlled by the cycle of volcanic activity and the deposition of calcium carbonate by inorganic chemical processes.

A normal atmosphere?

Summed up in one sentence, our atmosphere is a thin layer of inert nitrogen mixed with a highly active component – oxygen – with water and carbon continually cycling between the air and the planet itself. We may think of all these features as being "normal", and indeed most seem to be shared by the majority of worlds encountered in science-fiction ("you can remove your helmet, Captain, the air is breathable"). However, science has shown that no other atmosphere in the Solar System contains free oxygen. On Venus global winds constantly push sulphur clouds eastwards across the planet, and on Mars pole-to-pole storms of dust engulf the whole planet. Instead of a water or carbon cycle, Venus has two sulphur cycles, whereas Titan sports a methane cycle and a carbohydrate cycle.

The ages of Earth

Even the Earth's atmosphere hasn't always been this way, and it will not always be like this in the future, either. In fact, when we take a look at planet Earth at other times in its long history, it may look to us like a planet that is even more alien than Mars or Venus.

Geologists pieced together the past climates of the Earth from the study of ancient rocks and sediments. From a series of Ice Ages over the past three million years, to a much warmer planet in the time of the dinosaurs, the erratic blocks left behind by retreating glaciers and microscopic tropical shells buried in chalk cliffs tell of huge climatic variations. These changes were due to the rearrangements of the continents over the eras, to secular perturbation of the Earth's orbit, as well as to the slow depletion of carbon dioxide in the atmosphere by the rock-carbon cycle.

For instance, the Ice Ages were partly caused by the opening of the passage between South America and Antarctica, which locked the Antarctic continent on itself in a belt of cold sea current and made it become much cooler. With a cooler and whiter Antarctica (white reflects more of the sunlight back into space), the whole planet cooled. The tilt of the Earth nudged by the influence of other planets also played a role, slightly reducing the amount of light reaching high latitudes. Another factor in causing the Ice Ages was the rise of the Tibetan plateau following the collision of the

---

[5] The proven global oil reserves are about 200 billion tonnes, or about half the total present biomass.

Indian and Asian continental plates; this then created a Monsoon climate over South Asia, reducing the greenhouse effect by increasing the sequestration of carbon dioxide from the atmosphere.

Further in the past, 2.2 billion years ago, the Earth looked like a very different planet. The whole globe was plunged in an Ice Age much deeper than the one endured by the mammoths and Neanderthals, with most of the ocean and continents entirely covered in ice – one of the *Snowball Earth* episodes.

Three billion years ago, the atmosphere was entirely oxygen-free. Unprotected by the ozone layer, the continents were endless deserts of rocks sterilised by the Sun's hard, ultraviolet rays. Life was already abundant in the oceans though, mainly under the form of vast pools of blue-green algae, sheltered from ultraviolet light by water.

Closer still to the birth of the Earth and the Solar System, 4.6 billion years ago, our planet would have been a hellish place, with endless plains of recent lava flows and freshly-formed crusts over the magma, shrouded in a dense fog of water vapour and carbon dioxide, and with abundant amounts of sulphuric acid thrown in for good measure.

We shall visit these alien planets called Early Earth at the end of this book. Let us now embark on our tour of the full diversity of planetary atmospheres. First, let's meet our closest neighbour.

Chapter 2
Venus

Our planet has a sister called Venus. At birth they were almost identical twins, very closely matched in weight, size and composition – a molten iron core with a rock mantle, and a thick atmosphere of water and carbon dioxide. Since then, their experiences and the distant orbits in which they ended up made them mature into very different planets.

Most of the differences are due to one overwhelming factor – Venus is 30 percent closer to the Sun and receives about twice the amount of sunlight that reaches the Earth.

What does it do to a planet to receive twice as much light? As far as the bulk of the planet is concerned, not much, but in the atmosphere the changes are dramatic. We Earthlings know what such a change feels like, because similar variations happen here between polar, temperate and tropical regions.

The amount of light reaching Siberia or Alaska is half that of equatorial regions like the Amazon or Kenya – simply because the ground is inclined respective to the Sun, so that the same amount of sunlight is spread over more surface.

Fig. 2.1 The British Isles and Egypt as seen by the Sun in summer (left) and winter.

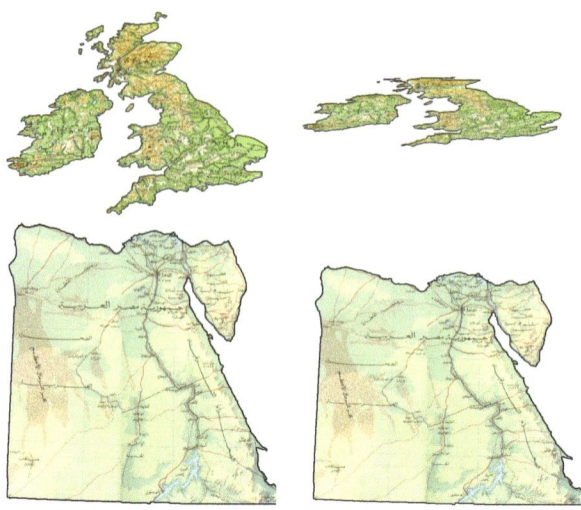

On Venus, regions close to the poles get as much sunshine as the Amazon, and equatorial regions are boiling hot. The direct sunlight at noon would bring a plate of metal up to 200 degrees Celsius. This would raise temperature at the surface of the planet above the boiling point of water, and over time any water ocean on Venus would entirely evaporate and fill the atmosphere with vapour.

One could imagine that Earth's twin sister would end up with an extremely thick and hot water vapour atmosphere, which some scientists have called a "Sauna planet". (In fact "Steam planet" would be more accurate, since saunas are very dry places. A hot water vapour atmosphere would be more akin to the interior of a steam engine.) This is not, however, what was found by the Soviet Venera probes that landed on Venus.

Before the Soviet and American interplanetary missions of the 1970s, observations of Venus from the ground had revealed that the planet was entirely shrouded in thick clouds and that these clouds were rotating around the whole planet in just three days. Measurements of infrared light showed that the temperature at the top of the clouds was a comfortable +10 degrees Celsius, compared to temperatures around – 60 degrees Celsius at the top of high clouds on Earth. This information naturally gave rise to the hypothesis that the planet was a warmer, wetter version of Earth.

It was only a small step, happily taken by many distinguished astronomers, to postulate that Venus was an equatorial paradise teeming with life, with dripping giant purple trees and distant sounds from numerous species of weird, unknown wildlife. The golden age of science fiction in the mid-twentieth century thrived on such speculation.

After 1945, new measurements in the infrared, microwave and radio wavelengths revealed that the air of Venus was composed mostly of carbon dioxide, with no trace of oxygen, and that when gaps occasionally appeared in the upper clouds, the ground beneath seemed to show alarmingly high temperatures of up to 400 degrees Celsius.

Nevertheless, when the flotilla of cold-war probes approached the shy planet, visions of alien rainforests were still very vivid.

A total of ten Soviet probes have landed on Venus, and thirteen orbiters have sent back data on its atmosphere. Because of the no-nonsense, hard-line approach to science of the former Soviet Union, the Venera probes brought back few pictures, but they did measure the temperature, winds and air composition on the planet.

Fig. 2.2 One of the splendidly Sovietic Venera probes. Image credit: NASA

The atmosphere of Venus is very thick, ninety times heavier than that of the Earth. This is enough pressure to crush a modern nuclear submarine. The whole planet is shrouded with thick cloud that culminates about 70 kilometres (40 miles) above the surface, and extends down to 48 kilometres (30 miles). There is practically no water in the clouds – the whole atmosphere is extremely dry, with less than 0.002 percent of water overall. The air is almost entirely made of carbon dioxide. The clouds are made of droplets of sulphuric acid. If the temperature is pleasant at the top of the clouds, it rises steadily underneath, until it reaches about 500 degrees Celsius at the surface.

If Earth is heaven (arguable but debatable), Venus is very close to hell.

Fig. 2.3 Returning Chinese astronauts. Each nation brings some of its heritage to the aesthetics of space exploration. What will Chinese probes look like? Image credit: Xinhua

Size	12,104 km (7,521 miles)	12,756 km (7,926 miles)
Gravity	0.92	1
Pressure at surface	92	1
Temperature at surface	480 C	20 C
Total water	10 cm	3 km

Earth   Venus   Earth   Venus

Fig. 2.4 (i-iv) Earth and Venus with their clouds off – and on. Image credit: NASA

Turning Earth into Venus

Scientists think that the atmospheres of Earth and Venus were rather similar after the birth of the Solar System, and that their divergent fate over the following 4.6 billion years was due to their different distances from the Sun. To try to understand how they became so different, let us imagine what would happen if our planet was moved closer to the Sun, to orbit one hundred million kilometres from the Sun instead of one hundred and fifty million.

Fig. 2.5 Size of the Sun from Earth and from Venus.

In the far North, the increased sunshine renders the climate of places like Alaska and Siberia much more pleasant. The mean ground temperature climbs up to 30 degrees Celsius off the northern shore of Siberia. More water evaporates from the sea and lakes into the warmer air, and big tropical thunderstorms refresh the taiga.

Near the tropics, meanwhile, the change is not so benign. The main driver for weather on Earth is the evaporation of water and lifting of the air by heat in the equatorial region. The added sunlight boosts this cycle, making the weather more extreme. The number and strength of tropical hurricanes increases steeply for each additional fraction of a degree added to the global temperature. With twice the amount of sunshine, the change is incomparably greater, with catastrophic hurricanes becoming as frequent as summer storms are today.

Rapidly, the level where clouds and storms stop moves higher than its present 10–15 kilometres because more heat is welling up from the ground.

As the temperature near the Equator reaches 60 degrees Celsius, one unfortunate consequence is that forests and animals start dying out. Plants and animals cannot survive prolonged periods of such heat.

The speed at which the climate heats up depends primarily on the proximity to the oceans. In continental deserts such as the Sahara, the change takes only a few hours. But it takes more than a century to heat the oceans entirely, so in coastal regions and islands the change is much more gradual, giving enough time for mobile creatures to fly, run or hop towards the polar regions.

In the polar regions, conditions become pleasant enough for most species, including humans. Deep in the oceans, the changes are not felt for years; luminescent fish and blind crabs quietly go about their business, marvelling at the increased amount of nutrients drifting down from above.

It takes only a few years for all remaining glaciers and polar ice caps to melt away, increasing the levels of the sea by 80 metres (260 feet), engulfing the deserted remains of most of the world's major cities.

About a century later, when the oceans have finally become warm throughout, the air temperature reaches 100 degrees Celsius near the Equator. Then equatorial oceans start boiling away, filling the air with huge clouds, and drowning higher latitudes with diluvian rains.

At that point Earth would be about 100 degrees Celsius warmer on average because of the increased sunshine – still a long way from the ground temperature on Venus. Imagine a world where tropical regions are sizzling deserts with boiling oceans, with a few more temperate patches remaining near the poles or at the top of the highest mountains, and life still teeming at the bottom of the warm polar oceans.

Greenhouse to madhouse

Just as the world tries to adapt to these new, warmer conditions, it gets much worse. The greenhouse effect of water vapour sends the atmosphere into ever-increasing temperatures. Water vapour is a greenhouse gas like carbon dioxide, and as the greenhouse effect of the added vapour kicks in, the temperature rises. In the greenhouse effect (the lower glass pane of Chapter 1), as more carbon dioxide is pumped into the air, more sunlight gets trapped near the surface. The problem with water vapour is that as the temperature rises, more water evaporates from the oceans. This is a positive feedback loop, a vicious circle in plain English: more water evaporates, the greenhouse effect becomes stronger, the temperature rises, more water evaporates.

On Earth, the effect of water vapour also amplifies the greenhouse effect of carbon dioxide, but the feedback loop stays under control. However, in the case of our overheating Earth, the feedback loop goes wild. The heat increases the evaporation of the oceans, exacerbating the greenhouse effect, and there is no limit until all the water in the ocean is vaporised.

At some point there is more water than nitrogen and oxygen in the atmosphere. The atmosphere as a whole becomes thicker. When the first 100 metres of the oceans have boiled away, the atmosphere is 90 percent water, and the pressure at the surface is 10 bars. At this point, the surface of the ocean is simmering at around 150 degrees Celsius – hotter than 100 degrees because the boiling point of water rises with the higher pressure.

Earth has now become a giant steam chamber. The whole planet is shrouded in thick clouds. Above the clouds, the temperature is relatively cool – below 20 degrees. Raindrops sometimes form, but evaporate before reaching the ground. Only in some remote corners of the Arctic and on high mountains does warm rain occasionally reach the ground. Steaming, temporary rivulets snake down the slopes, but evaporate before reaching the boiling sea. The last human survivors gather underground near these streams, in volcanic caves under Mount Erebus in Antarctica.

Since the weight of water in the ocean is 300 times larger than the total weight of the air in the atmosphere, when the last puddle of the former global oceans vanishes, the atmosphere is 300 times heavier than now, and made up of 99.7 percent water vapour. At that point, a few decades after bringing it nearer to the Sun, the planet looks like Venus in many respects, entirely shrouded in thick clouds, with an enormous surface pressure and very high temperature.

Something very strange happens by the time the oceans are entirely vaporised – something that brings out an aspect of the nature of water that our intuition has not prepared us for (as we'll see in Chapter 6, there are more surprises in store from that apparently familiar substance). The pressure and temperature now exceed the so-called "triple point" of water, at 221 bars and 374 degrees Celsius, the point at which liquid water and vapour become indistinguishable. This is known as "super-critical water" and is difficult for us to imagine. The only place where water takes such a form naturally on Earth is at the mouth of undersea volcanoes.

Under normal pressure, when water vapour cools from, say, 500 to 50 degrees Celsius, it abruptly condenses from a gas to a liquid when it crosses the 100 degree mark.

However, under very high pressure, anywhere above the critical point at 221 bars, at 500 degrees Celsius water is a gas-like vapour, and at 50 degrees it is a liquid-like fluid, but there is no specific temperature at which an abrupt transition occurs. There is no longer any surface between atmosphere and ocean, the change is gradual, like honey or butter coalescing as it cools.

Water escape

There is still one key difference between this steam-chamber Earth and Venus: the atmosphere is made almost entirely of water rather than carbon dioxide. We now need to run the clock forwards for millions of years rather than decades to understand what happened to the water on Venus.

Water is one of the most abundant substances on the surface of planets, and Earth, Venus and Mars probably all possessed generous amounts of surface water in the early days of the Solar System. However, water has a fundamental weakness – as H_2O, it contains two atoms of hydrogen, the lightest element in the Universe. The gravity of Earth is not strong enough to keep hold of free hydrogen. As long as the hydrogen atoms are tied up in a water molecule, they are fine, but if they ever become separated, they drift off into space.

That is why there is no free hydrogen in the atmosphere of Venus, Earth or Mars, despite the fact that hydrogen is by far the most abundant element in the universe and makes up most of the Sun, Jupiter and Saturn. Giant planets and stars are heavy enough to keep hold of all atoms, including hydrogen, but not Earth-mass planets. Their gravity can only retain heavier molecules, like water (nine times the weight of hydrogen), nitrogen (14 times) and carbon dioxide (22 times). At the other end of the scale, small celestial bodies like the Moon cannot even retain these relatively heavy gases, and soon become airless.

Every now and then, a water molecule in the air is broken apart by sunlight (ultra-violet light can do that), and the two hydrogen atoms can escape into space. Slowly but surely, the hydrogen in the water leaks away and, over millions of years, all of the hydrogen in the water of a planet becomes lost to space. The oxygen remains behind. Since oxygen is a very reactive element, it usually combines with carbon to form CO_2 or with rocks to form oxides.

Earth has not lost its oceans because the temperature in the upper atmosphere is low enough that water condenses before it reaches the altitudes at which it could be dissociated by sunlight and escape forever. With the atmosphere heated by more intense sunlight and an enormous greenhouse effect, the water vapour in our hotter Earth leaks out into space inexorably.

Volcanic gases

On Earth, dozens of volcanoes erupt every year, and they range from discreet smoke plumes to colossal disasters that can change the climate of the whole planet. The Tambora in Indonesia caused the "year without a summer" in Europe in 1816, and Santorini wiped out the Minoan civilization of Crete during the Bronze Age. Volcanoes spit out a vast array of gases, but the dominant ones are carbon dioxide and sulphur. In our planet, these gases do not accumulate in the atmosphere because they are recycled into the oceans or the ground; carbon dioxide is fixed into carbonate rocks, while sulphur dissolves in rain drops, producing "acid rain" that ends up in the oceans, ultimately finding its way back into the ground through ocean sediments.

However, both these processes require water. With the water gone on our hot Earth, carbon dioxide and sulphur simply accumulate in the air. Over millions of years, as the plumes of every volcano that has ever erupted merge into a single, planet-circling cloud, the whole atmosphere is choked in carbon dioxide and sulphur. There is about 40 atmospheres worth of carbon in the ground on Earth, most of which has cycled through the atmosphere at some point. If this carbon accumulates in the atmosphere, the pressure at the surface of the planet will exceed 40 bars – the pressure at the bottom of an imaginary pit some 20 miles below ground.

The temperature at the surface rises to several hundred degrees, under the combined effect of increased pressure and the powerful greenhouse potential of CO_2. The air is mostly carbon dioxide, with thick, yellowish clouds of sulphuric acid hovering in the haze. Sulphuric acid rain constantly falls, but evaporates before reaching the ground. The landscape of red-hot rocks shimmers like a mirage in the heat. We have reached our destination, the planet Venus.

The real Venus

A ground of hot volcanic rock; air a poisonous mixture of volcanic gases at high pressure, so thick that it feels as much like a liquid as a gas, gloomy red light filtering through thick clouds. The closest Earth analogue to these nightmarish conditions at the surface of Venus may be sitting on top of a deep-ocean volcano.

Fig. 2.6 View of the surface of Venus from one of the Venera probes. The geometry of the image has been reconstructed and the colours adapted to make them as close as possible to what a human eye would see. Image credit: Don Mitchell

Fig. 2.7 One of the original images from Venera-13. Image credit: NASA

Fifty kilometres above the inferno float the gloomy sulphuric-acid clouds, making the air as acidic as pure lemon juice. About seventy kilometres (40 miles) above the surface, the sulphur clouds start to break and the Sun becomes visible at last, albeit through a yellow haze that extends a further twenty kilometres upwards.

Venus is a highly volcanic world, its surface is littered with recently solidified lava flows and volcanic domes. Volcanoes exist because planets the size of Venus and Earth have a lot of internal heat to evacuate. Planets are born from millions of collisions that leave their cores at thousands of degrees, and when their exterior is made of solid rocks, the only way to evacuate enough heat is through episodic eruptions of lava and gas.

The volcanic nature of Earth is partly concealed by the effects of the oceans and the water cycle. Most volcanic activity occurs underwater, at the mid-ocean ridges where crustal plates are constantly re-generated, and at the boundaries between tectonic plates.

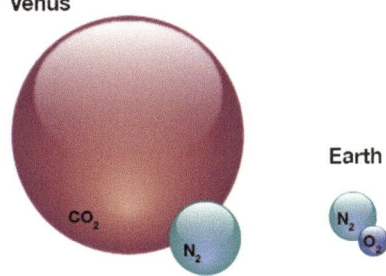

Fig. 2.8 The composition of the atmosphere of Venus (compared to Earth on a scale proportional to the total quantities) mostly carbon dioxide, with a few percent of nitrogen (still amounting to three times the total amount in our atmosphere) and a drizzle of sulphuric acid.

Oceanic lava flows are quickly covered by sediment, and on land volcanic features are erased by erosion and covered by vegetation. Although traces of volcanism are hard to find on Earth, over geological timescales volcanism is one of the main actors in the drama of the Earth's surface and atmosphere. A large fraction of the present land masses consist of volcanic outpourings of magma, such as the Siberian Traps in Russia, the Deccan Traps in India, or the Columbian River Basin in the United States.

Volcanoes on Venus are different in aspect from Earth's volcanoes. Because of the higher surface temperature and absence of water, Venusian eruptions rarely exhibit the cone shape common on Earth, they form mushrooming blobs or meandering lava flows.

The clouds of Venus

Comparing conditions at the surface of two planets can be misleading. When describing a position within an atmosphere, a more natural coordinate than altitude is pressure, which really tells what the place feels like. The drawing overleaf shows the atmospheres of Earth and Venus compared on the same pressure scale. From that viewpoint, the surface of Venus corresponds to the sea floor on Earth, and our sea level equates to about 50 kilometres (30 miles) above ground on Venus, where both the pressure and temperature are similar to those we are used to.

Abandoning our parochial attachment to solid ground, we could arguably call the cloud-tops the "surface" of Venus, in the same way that we consider the top of the ocean to belong to the surface of the Earth, rather than the dark sea floor miles underneath. The thick clouds shroud the entire planet continuously, and that is what we see from Earth. In the Earth-like conditions of pressure and temperature at this level, many chemical reactions can take place in the droplets of sulphuric acid. Some Venus specialists even suggest that we should be checking these clouds for signs of life, since the sulphur chemistry may be complex enough to give rise to biological processes.

Superficially, the clouds on Venus are reminiscent of clouds on Earth but in reality they are extremely different. There is the lemon-juice acidity, but also the fact that whereas clouds on Earth form and dissolve in matters of hours or days, those of Venus ebb and swell but never break. They are also much thinner than Earth's clouds, more like a morning fog or, indeed, industrial smog over a city. The visibility within Venusian clouds is several kilometres, as opposed to a few metres in our water clouds. The Venusian cloud cover is opaque not because the clouds are dense, but because they extend over such a deep layer. Venusian clouds are more transparent than ours because the grains and droplets in them are much smaller, a few micrometres across, whereas

1 mbar

10 mbar

100 mbar

1 bar

10 bars

100 bars

sulphur haze

20° C
upper clouds
+70 km

lower clouds

sulphuric
acid rain

480° C

Fig. 2.9 Profile of the atmosphere of Venus and Earth, on the same pressure scale.

Fig. 2.10 Venus in real colours.
Image credit: NASA

typical ice crystals in our clouds measure hundreds of microns, and snowflakes or hail stones can be much bigger. Venus's clouds are also more fluffy, so it is difficult to tell where they begin or end. More than anything, they are far larger than any clouds we know.

The clouds of Venus extend over nearly 30 kilometres (20 miles) in height, in three different layers separated by intervals of clear air. The upper layer is a gradually thickening haze, exposed to full sunlight during the day. In the middle is the main cloud deck of solid and liquid grains, drifting down and raining until they vaporise. The lower deck is formed by upwelling currents which cool sulphuric acid to temperatures low enough for it to condense, in the same way as cumulus clouds form on Earth.

The role of sulphur

Sulphur dominates the chemistry of the atmosphere of Venus. Chemically, sulphur is a very versatile atom that shares some properties both with oxygen and carbon. Like oxygen, it has a propensity to bond strongly with most other elements. Like carbon, it can also bond to itself in chains (although to a lesser degree than carbon). In an oxygen-poor environment, sulphur combines with hydrogen to form hydrogen sulphide (H_2S), the nasty, explosive gas that gives a rotten-egg smell to hot-spring water. If carbon and oxygen are present, sulphur can replace the carbon atom in CO_2 to form SO_2, sulphur dioxide, then H_2SO_4, sulphuric acid.

Sulphur has a very bad reputation on our planet. It is "brimstone" in the Apocalypse of Saint John, and has been called the element of the Devil, one reason being the association of sulphur with volcanoes. Apart from the rotten egg smell, sulphur compounds are also behind the strong flavour of garlic and onion. Elemental sulphur is a yellow powder, rarely found in that form except when it falls from the sky with ashes and cinders after a volcanic eruption.

The clouds of Venus are part of the sulphur cycle, analogous to the formation of acid rain from pollution on Earth. The sulphur dioxide in the lower atmosphere is dragged up by convection, and when it surfaces above the clouds it can react in the sunlight to form sulphuric acid. As the temperature in the upper atmosphere is low enough for sulphuric

Fig. 2.11 Sulphur haze.

acid to condense it will form tiny droplets, making the atmosphere hazy just like smog over a city. Over time the largest drops will start falling, growing in size to form the clouds below. At some point however, the temperature will reach the boiling point of sulphur dioxide (337 degrees Celsius), and the rain will evaporate, never reaching the ground.

Finally, in the high temperatures below the clouds, the sulphuric acid is broken back into its constituent elements, including smelly sulphur dioxide, ready to be carried upwards by the convection of the atmosphere for another cycle.

Fig. 2.12 Acid rain on Earth. Sulphur produced by industries – or by volcanoes – is converted to sulphuric acid by contact with water. The acid is carried with the rain, and returned to the ground.
Image credit: Joanna Barstow

There is more to the sulphur cycle than the "acid rain loop". Chemical reactions triggered by sunlight in the upper clouds can also produce chains of sulphur atoms that may change the colours of the clouds and give them a yellow tinge. Sulphur also reacts with the rocks on the ground. Like water on Earth, sulphur cycles between solid, liquid and gas, but in addition, it undergoes chemical changes along the cycle; in that sense the Venusian sulphur cycle is more like the carbon cycle on Earth. In fact, sulphur does both cycles at once. The "fast" cycle between the lower atmosphere, the upper haze and the clouds, is analogous to the production of acid rain on Earth, and takes place in hours or days. The "slow" cycle between the atmosphere, the rocks and the volcanoes, resembles the rock-carbon cycle, and takes years.

Fig. 2.13 The sulphur cycle(s) on Venus. Image credit: Joanna Barstow

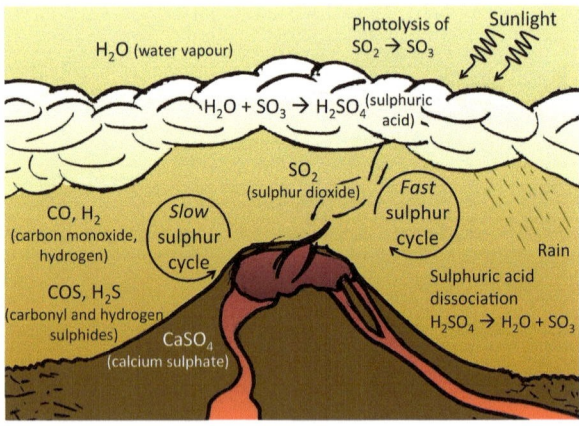

Super-rotation

The clouds of Venus spin around the whole planet every few days. Before they were able to observe the surface, astronomers assumed that this was the speed at which the planet turned on itself. This did not seem unreasonable compared to the 24-hour rotation period of Earth or the 25 hours on Mars.

But in fact, the planet Venus takes 243 days to spin on itself, or eight Earth months! Thus the surface hardly moves at all below the furiously blowing winds that carry the clouds around. The clouds always move from west to east, so it looks as if the planet is spinning on itself much faster than it actually is. The atmosphere is said to be in *super-rotation*. The speed of the winds that push the clouds around reach 300 kilometres per hour (200 miles per hour) in the upper cloud deck. They decrease to 200 kilometres per hour at the bottom of the clouds, and fall to a few kilometres per hour near the ground. However, since the atmosphere near the ground is so much denser, this "breeze" is still able to push dust and even pebbles around like a strong gale can on Earth. This is difficult for us to imagine – a gas so thick that the wind is almost like a current in water.

Why does the whole cloud system rotate around the planet from west to east, as if the planet was spinning on itself? The culprit is the Coriolis effect.

The dominant movement in the atmosphere of Venus is, like on Earth, the circulation between the equatorial regions and the middle latitudes. But since the planet is rotating much more slowly than Earth, the Equator-to-pole rolls get much closer to the poles before being deflected by the Coriolis effect. This time instead of imagining walking on a spinning platform like in Chapter 1, we can think of the "ice skater" illustration of the effect. Consider a ring of clouds that slowly moves round Venus near the equatorial regions, and is brought polewards by the heat circulation. As the whole ring moves polewards it becomes more compact, like a skater bringing their arms closer to the body, which will cause it to spin faster.

In this way the polewards motion of air creates a global eastward rotation, and integrated over the whole planet it gradually makes the whole atmosphere spin in the

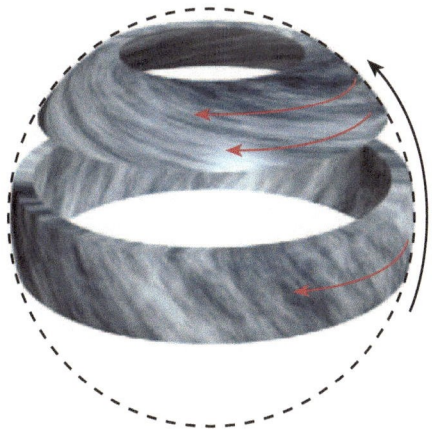

Fig. 2.14 (a/b) Ice skaters spin up by gathering their limbs closer to their axis of rotation. A circular atmospheric current moving towards the pole of a planet spins up for the same reason.

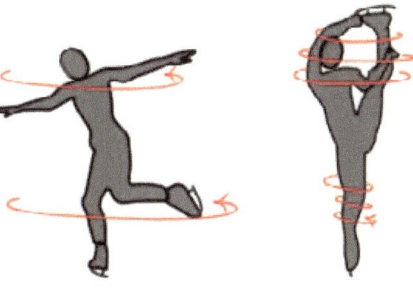

direction of the rotation. Even the slow rotation of Venus (243 days versus one day for Earth) is sufficient to get the eastward flow going, and, once it is set, it keeps being accelerated until the clouds rotate much faster than the planet. Indeed, although we have not yet observed other Earth-like exoplanet atmospheres, the same phenomenon can be seen for hot Jupiters; their atmospheres circle in a few days at most, regardless of the spin of the planet itself. At the kind of temperatures found at the top of the clouds of Venus or on Earth, it takes a few days for 'air' to lose its heat by radiation into space. So, in order for the atmosphere to carry the heat around the planet, it needs to travel from the day side to the night side in less than a few days, otherwise it will have lost its heat. On Earth, the air can hitch a ride on the 24-hour rotation of the planet, so winds are not needed to transport heat from the day side to the night side[6]. But if the planet spins more slowly, the atmosphere as a whole will rotate in a few days to be able to transport the heat from west to east. To an outside observer who cannot see through the clouds, the planet will appear to be rotating in a few days.

This has an implication for the study of Earth-like planets around other stars: from the outside they can appear to be rotating with periods of less than a few days, even if the bulk of the planet is rotating more slowly, or not at all.

Being there

What would it feel like to stand on Venus? At first, it would be like standing inside a deep-sea submersible such as the Russian-designed MIR that filmed the remains of the Titanic. You need this kind of vehicle to resist the 90-bar pressure. Instead of a wrecked ship, one would be peering at a desert landscape in a reddish glow; until the heat kicked in. Then it would feel like being stuck in an oven, with uncomfortable consequences.

The surface of Venus has limited touristic potential. The conditions are much more amenable at the cloud-top level, where, equipped with a proper hot air balloon, it would be possible to enjoy normal atmospheric pressure and pleasant temperatures around 20 degrees Celsius. The problem there would be the vitriolic fog, with the acidity of the clouds high enough to dissolve skin. Nothing, however, that a rubber suit couldn't stand.

Overall, a perfectly decent way to spend an afternoon.

---

[6] The winds nevertheless take charge of the Equator-to-pole heat transport.

Fig. 2.15 Ballooning in the upper cloud deck of Venus.

Chapter 3
Mars

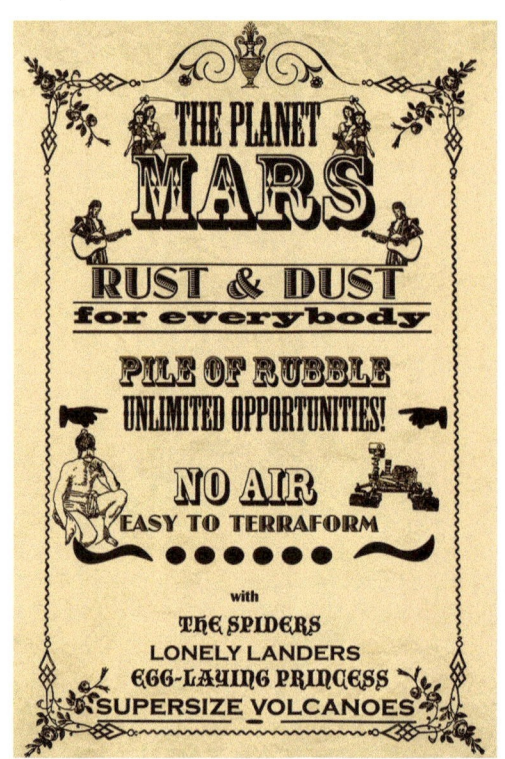

Little brother

Earth and Venus have a little brother – cold, barren and hostile… it is everyone's favourite. Mars has held a powerful grip on our imagination, ever since the Italian scientist Giovanni Schiapparelli used the word "canali" in 1877 to describe the features he glimpsed on Mars. A generation later, Percival Lowell "improved" these into irrigation canals, cities and dying civilisations. The fascination with the Red Planet is still very much alive today, with the hard-to-resist urge to fill the dry or dusty ravines of the desolate planet, which we see in images sent back by the exploration rovers, with once-rich ecosytems or ancient civilisations.

Fig. 3.1 The magnificent real Mars. Spot the three giant volcanoes (along the bottom), the Marineris Canyon (running vertically), the Ares flow channels (the mouth of the Canyon), the rusty colour of iron oxides and the faint veil of dry-ice clouds (whitish patches). Image credit: NASA

Mars is a much smaller planet than Earth and Venus, about half their size, and a tenth of their mass. It was prevented from becoming an Earth-size world by the disrupting influence of its far more powerful neighbour, the giant planet Jupiter, which scattered the orbits of the planet-building asteroids in the area as the inner planets formed.

Mars orbits at 1.5 astronomical units, that is 50 percent further away from the Sun than Earth, and almost twice as far as Venus.

Fig. 3.2 Mars is about half the size of the Earth and a tenth of its mass; its atmosphere is one hundred times lighter. Image credit: NASA

Earth versus Mars

	Mars	Earth
Size	6,780 km (4,212 miles)	12,756 km (7,926 miles)
Gravity	0.4	1
Pressure at surface	0.008	1
Temperature at surface	-130 – +10 C	-20 – +50 C
Total water	10 cm	3 km
Distance from Sun	1.5 AU	1 AU

Fig. 3.3 The size of the rising or setting Sun from Venus, Earth and Mars (right).

The surface of Mars is a cold and arid desert of oxidised rocks (its ground is "rusty", hence the red colour). The temperature varies between 130 below zero and a few degrees above zero, depending on the latitude, altitude and seasons. At noon near the Equator it can briefly reach above freezing, while in the winter near the poles, it gets cold enough to freeze the air itself. The atmosphere is about one hundred times thinner than on Earth, consisting almost entirely of carbon dioxide. As regards temperature and pressure, the conditions on the surface of Mars are similar to those at an altitude of 50 kilometres above Earth. On the calderas of the giant volcanoes, towering some 25 kilometres (15 miles) above the surrounding plains, there is almost no air at all.

Two modifications would be necessary to transform Earth into a Mars-like planet: bring it further away from the Sun, and make it lighter.

Marsaforming Earth: moving away from the Sun

Let us take the Earth and abruptly move it away from the Sun, out to the orbital distance of Mars. Seen from there, the Sun is 60 percent weaker, which means that equatorial regions would receive as much sunlight as Siberia or Alaska today. A loss of 60 percent sunlight causes the global temperature of the planet to decrease by about 45 degrees Celsius.

How rapidly the temperature drops with the decreased sunlight depends mainly on the amount of humidity in the air. In arid areas the temperature will plummet mercilessly, just like it does every evening today in the Sahara desert. What little water vapour was in the air condenses and covers the dunes and rocks with a thin layer of frost. In wetter climes the change is gentler, and takes weeks to register – as does the onset of summer or winter in places like Britain, where the warmest and coldest times of year are offset by about two months from the times of highest and lowest sunlight. It is slowest in the oceans, where the amount of heat stored in the water is so great that it takes centuries to reach a new equilibrium when conditions change at their surface.

A few days after having moved the Earth away from the Sun, the temperatures are below freezing over most of the continents, and as the oceans cool, evaporation subsides and the water cycle slowly reduces to a trickle. Most of the water vapour in the atmosphere crystallizes and snows down to the ground. This takes place in the coldest regions first. The ocean ice caps creep out towards the tropics from both poles.

With less heat to push the air around, the whole climate of the planet becomes more benign. There is no longer enough heat for hurricanes or even summer cumulus clouds to form. The stratosphere moves lower, hovering only a few kilometres above the surface and the only clouds left are low, thin stratus (this is the climate of Antarctica today).

Two effects now amplify the drop in temperature: first, water vapour is a very strong greenhouse gas, and with most of the water frozen out of the atmosphere, Earth loses its blanket of greenhouse-effect heating and about 20 degrees worth of insulation; secondly, because snow is white and bright, a lot more sunlight now gets reflected straight back into space, rather than being absorbed by the green-brown continents or the dark blue sea. This has the same effect as pushing the planet even further out from the Sun.

On this cold Earth, people are only able to survive now in deep cave dwellings near the Equator, using solar panels and geothermal heat to melt ice and grow plants. As the ice caps close in on the last exposed puddle of sea water, the capacity of the oceans to warm the atmosphere is lost and temperatures crash all over the planet.

Near the poles, as the thermometer reaches the fateful threshold of –120 degrees Celsius, the air has become cold enough for carbon dioxide to start freezing. A new kind of cloud appears in the high atmosphere, thinner and higher than water clouds – CO_2 clouds. Dry ice, CO_2 snow, sometimes falls on top of the glaciers.

The Earth is now a frozen mass of ice and snow. In fact, we have good reason to believe that such a "snowball Earth" has existed in the past, at least once somewhere between 2.4 and 2.1 billion years ago, and maybe a few other times, including perhaps around 650 million years ago. We shall meet snowball Earth again in Chapter 7.

Marsforming Earth: making the planet smaller

In our imaginary experiment to transform Earth into Mars, we have moved the planet away from the Sun. The second step is now to shrink the planet to about half the size of present Earth and one-tenth of its mass. As before, we follow the dire consequences on the atmosphere.

Our poor imaginary friends tending their tomatoes and mushrooms in their caves face a new crisis. Shrinking the planet to the size of Mars reduces the gravity to about one third of what it was. The effect on the atmosphere is to reduce the pressure everywhere to a third. At first, this will simply cause the atmosphere to expand. As the pressure is lowered, the bulk of the atmosphere inflates to occupy about three times more volume, extending out to more than 300 kilometres above the surface instead of the present 100 kilometres. For humans this will pose an immediate problem; the air pressure at sea level has become that of the top of Mount Everest today. This is too low for us to adapt to, and our imaginary colonies must now take refuge at the bottom of deep mine shafts. Shifts of people with breathing equipment will come to the surface from time to time to tend the greenhouses. The only good news is that the deeper dwellings will be warmer than conditions on the surface.

In the longer term, however, lower gravity will spell doom, not just for the colony but for the whole atmosphere. Earth and Venus are too light-weighted to keep hold of hydrogen and helium, the lightest gases, which escape into space. At one-third of standard Earth gravity, it becomes difficult to cling on to the rest of the mid-weight molecules that make up the atmosphere, such as nitrogen and carbon dioxide. With the low gravity on Mars even these molecules will occasionally be tempted to leap out into deep space. A trickle, but steady enough for the atmosphere to be entirely lost by the time the planet reaches the present age of the Solar System. In a Martian earth the atmosphere will slowly but surely leak into space!

Fortunately the cold surface does not shut down the volcanoes, which will keep on releasing gas into the atmosphere. This is a slow process, occurring over millions of years, but so is the loss of atmospheric gases into space. The balance between gases coming out of volcanoes and those leaking into space will set the amount of atmosphere our Martian planet will be left with, as well as its composition. As on Venus, it will mostly be carbon dioxide (CO_2), because it is the main volatile constituent of volcanic fumes. With most of the atmosphere lost to space and only a trickle of volcanic gases to replenish it, the pressure at the surface of our planet is now only a hundredth of Earth's atmosphere.

Now that most of the air is carbon dioxide, the occasional dip below the minus 120 degrees Celsius threshold has stunning consequences. Carbon dioxide freezes at –120 degrees Celsius, but what happens is no ordinary snowfall, the atmosphere itself drops to the ground. It is as if air itself was "raining" or "snowing". If the temperature kept dropping, the whole atmosphere would end up as a thin coating on the planet, an icy, airless world, like the moons of Jupiter. Given the distance of Mars to the Sun, however, the sunlight still provides sufficient heat to keep some of the CO_2 in the atmosphere.

Seasonal changes on Mars are stronger than on Earth, because Mars is more inclined on its orbit (25 degrees instead of 23 degrees) and the Martian year is longer (668 days instead of 365). The South pole gets very cold in the southern winter as the North pole in the northern winter. Temperatures dip below the freezing point of carbon dioxide, so the main constituent of the atmosphere starts to freeze out and fall as dry ice snow on the polar caps. Nothing comparable happens on Earth; water can condense into rain and snow, but water always represents a small proportion of the air's composition.

The real Mars
Pressure between 0.3 and 11 millibars, composition 95 percent carbon dioxide, the rest

Fig. 3.4 Composition and size of the Martian atmosphere compared to Earth.

Earth   # Mars

mostly nitrogen and argon, temperature between –120 and 0 degrees Celsius, these are the life statistics of the present atmosphere on Mars: not much of an atmosphere to be honest, with a surface pressure only one hundredth of that at sea level on Earth today. That is 99 percent of the way to empty space. It may not be too obvious from science-fiction novels and NASA coverage of Mars exploration, but the atmosphere on the surface of Mars is much thinner than that on Earth even at five times the height of Mount Everest. The same spacesuits used for the Moon landings will be required for people to walk on Mars. At the top of Mars' giant volcanoes, pressure is so low that it is close to the definition of "space".

Yet, thin as it is, the atmosphere of Mars has its own clouds, winds and storms. The thin atmosphere of Mars has a salmon-coloured tinge. Like our atmosphere, it

Fig. 3.5 Profiles of Mars and Earth on the same pressure scale. The atmosphere of Mars is almost one hundred times lighter than on Earth, and the pressure at its lowest point – at the bottom of the Hellas giant crater – is only 0.01 bar. The mean temperature in the Martian plains is around –60 degrees Celsius, quite like that at similar pressures on Earth. At the top of the giant volcanoes the temperature plummets to –120 degrees Celsius. Carbon dioxide clouds are common in the salmon-coloured sky, and there are occasional water clouds. Dust storms gather near the surface at the warmest spots, which move with the seasons.

contains some blue coloration because all molecules in their gas form tend to diffract blue light more easily than red light. But it also contains some red, because of the sand dust suspended in the air. The result is an undefined and very Martian mix of colours, with subtle changes during the course of the day and between seasons. An added reason why the colour of the Martian sky is difficult to define is that it is quite dark, because the atmosphere is so thin and the Sun distant.

Occasional thin clouds can form in the atmosphere of Mars, with some snowfall, either normal water snow in the warmer places, or carbon dioxide snow in the cooler ones.

Fig. 3.6 Sunset from the Viking expedition landers. The colours of the Martian sky are subtle, with a reddish tinge because of dust, and a hint of blue some-times because of scattering by molecules; but mostly the colours are very weak because of the lack of air. Image credit: NASA

The main characteristic of the Martian climate though, is dust. With constant winds and no water to keep the sand down, the surface of Mars is a paradise for sand storms. Grains get endlessly grinded down until they are tinier than the thinnest sand of the whitest atoll on Earth, and can be picked up by the breeze. Martian mini-tornadoes roam the surface, leaving worm-like traces in the sand. Fully-fledged dust storms are more common when winds become stronger, and, from space at least, they look very similar to their Earth equivalents. Every few years, the dust dynamics on Mars become even more extreme, with the dust storms merging into a single, planet-wide event that shrouds the whole planet in a blanket of drifting clouds for months on end. During these global dust storms, only the tops of the highest volcanoes emerge from the haze, giving the planet a very different aspect.

Fig. 3.7 A dust storm on Earth. Image credit: NOAA/George E. Marsh

Figs. 3.8, 3.9 Dust storm on Earth (over the North Atlantic, the Sahara on the right, Spain in the upper right); dust storm on Mars, on the same scale. Strong winds are lifting the sand/dust in the air and they get carried along in the atmosphere, tracing atmospheric motions. Image credit: NASA

The role of dust

Dust affects the climate of a planet. Like industrial smog on Earth and the black smoke curling out from the coal factories of old, dust blocks out some of the incoming sunlight, keeping the surface cooler. In fact, since dust is more transparent to infrared radiation than to visible light, its effect is almost opposite that of greenhouse gases, and scientists sometimes talk about the "anti-greenhouse effect" of dust. Sunlight is absorbed or reflected back, and infrared light escapes, so that the regions below the dust remain cooler than they would be otherwise.

Fig. 3.10 The greenhouse effect versus the "greyhouse effect" of a dust cloud. In the classic greenhouse effect, the atmosphere is transparent to visible sunlight and opaque to infrared radiation ("heat"), increasing the temperature below. Dust and other fine particles can have the opposite effect; they block sunlight, and infrared light leaks through them. This is the mechanism of volcanic cooling and the feared "nuclear winter" scenario.

In 1997, when the danger of the CO_2 greenhouse entered global consciousness in the run-up to the Kyoto conference on global warming, climate models included the effect of the greenhouse gases, but the influence of dust was not considered in detail.

When it was subsequently included, something really paradoxical emerged: in the most industrialised regions of the world – Europe, China and the Eastern United States – the haze of fumes and soot from industrial pollution was serving as an anti-greenhouse blanket and locally cooling the climate. As a result the regions most responsible for greenhouse gas emissions were partially protected from climate change.

Fig. 3.11 A coal factory doing its bit for climate control by producing aerosols (if only it wasn't also pouring carbon dioxide into the air). Image credit: Adam Cohn

This suggested a possible if rather extreme way to fight global warming on Earth: by releasing enough soot, smog and pollution as high up as possible in the atmosphere. Of course, critics are quick to point out that this would come with unintended consequences (such as poisoning the oceans) likely to make the cure worse than the disease. On the other hand, this process is already happening naturally; volcanic eruptions can release huge amounts of ash into the atmosphere, sometimes enough to cool the climate over the whole planet. When Mount Tambora erupted in April 1815, dozens of cubic kilometres of stones and ash were thrown into the sky – the largest eruption in a millennium. The ash clouds took a year to settle to the ground, mainly via rainfall, and had such a strong effect on the climate that 1816 became known as the "year without a summer", with widespread crop failure worldwide. In 2009, inhabitants of northern Europe were reminded of the power of volcanoes when the Eyjafjallajökull in Iceland started spewing enough dust into the air to hamper air traffic on a continental scale.

It says something about the fragile nature of an atmosphere compared to the bulk of its planet that a single volcano, erupting at the right moment, could reverse global warming for years. Is a volcanic eruption going to be our best remaining hope of avoiding catastrophic climate change in the coming years? That would be some irony. The catch is that the short-term consequences of a large volcanic cloud, such as crop failures, would be a terrible price to pay.

Another way to fill the atmosphere with dust is to detonate a large number of thermonuclear bombs at ground level. In the 1960s, some scientists raised the alarm about the possibility of a "nuclear winter" following a large-scale nuclear war between the

United States and the Soviet Union. The millions of tonnes of fine dust thrown up into the upper atmosphere by nuclear blasts would darken the skies for months, crippling agriculture on a global scale.

The late Carl Sagan has run simulations of what a nuclear winter would be like, and has written several chilling accounts of the results:

In the baseline case, the amount of sunlight at the ground was reduced to a few percent of normal – much darker, in daylight, than in a heavy overcast sky and too dark for plants to make a living from photosynthesis. At least in the Northern Hemisphere, where the great preponderance of strategic targets lie, an unbroken and deadly gloom would persist for weeks.

Even more unexpected were the temperatures calculated. In the baseline case, land temperatures, except for narrow strips of coastline, dropped to minus 25 Celsius (minus 13 degrees Fahrenheit) and stayed below freezing for months – even for a summer war.

The oceans, a significant heat reservoir, would not freeze however, and a major Ice Age would probably not be triggered. But because the temperatures would drop so catastrophically, virtually all crops and farm animals, at least in the Northern Hemisphere, would be destroyed, as would most varieties of uncultivated or domesticated food supplies. Most of the human survivors would starve.

Martian climate

The climate of Mars is also affected by the way the planet rotates around our Sun. Mars is slightly more tilted on its orbit than Earth, 25.2 degrees versus 23.4, which makes the seasons more pronounced. Moreover, the orbit of Mars is quite ellipse-shaped (eccentric) because of the influence of Jupiter, which makes the seasons more marked in the Southern Hemisphere than in the Northern Hemisphere.

Surface features on Mars are vaster and larger than on Earth. This is partly because lower gravity allows for higher relief, and partly because of the history of large impacts during the early life of the Solar System. Most of the Southern Hemisphere is a highland plateau, raised 4 kilometres (2.5 miles) above the flatter lowlands of the Northern Hemisphere. Four giant volcanoes rise more than 25 kilometres (15 miles) above the surface, and a gigantic canyon as long as the continental United States crosses the equatorial regions.

Each winter, when the temperature dips low enough for carbon dioxide to condense, about a third of the atmosphere snows out on the ice caps. The atmospheric collapse sucks in fast winds from lower latitudes towards the poles, causing huge planet-wide storms as currents from the parts of the planet most exposed to sunlight rush in to replace the frozen atmosphere. Each spring the frozen atmosphere evaporates again.

Mars rotates on itself in almost 25 hours (24 hours 37 minutes 22 seconds to be precise). Although this value is uncannily close to an Earth day, it implies a much smaller Coriolis twist for currents on the planet for two reasons: firstly, for a smaller planet a given rotation period translates into a lower rotation speed at the surface; secondly, there is less distance to cover on the planet from Equator to pole.

As a result, atmospheric currents are able to cross the whole planet from pole to pole when the winds are strong enough. In particular when in the depth of winter the carbon dioxide atmosphere starts condensing and triggers the seasonal atmospheric collapse, warmer winds from the other hemisphere that rush in to fill the gap merge into a single pole-to-pole swirl. The result can be dramatic and even visible with a good telescope from Earth, since the winds pick up dust and can cover the whole planet.

Fig. 3.12 Mars before and during a global dust storm. Note in the first image the South Polar Cap and the clouds in the colder regions. Image credit: NASA

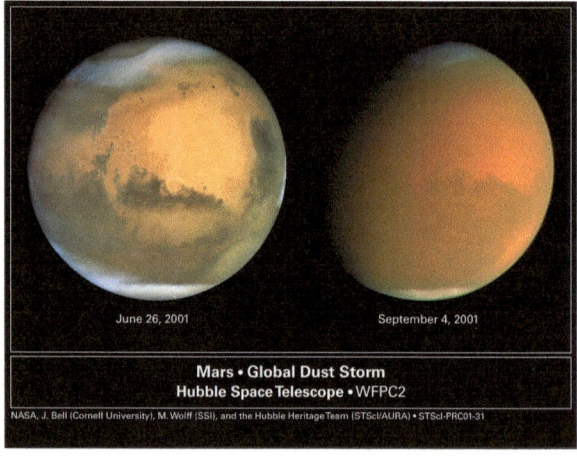

Planetary scientists have now gathered such a detailed knowledge of the Martian atmosphere that they can adapt the global models used to forecast the weather and predict the climate on Earth, and run them to understand the climate on Mars.

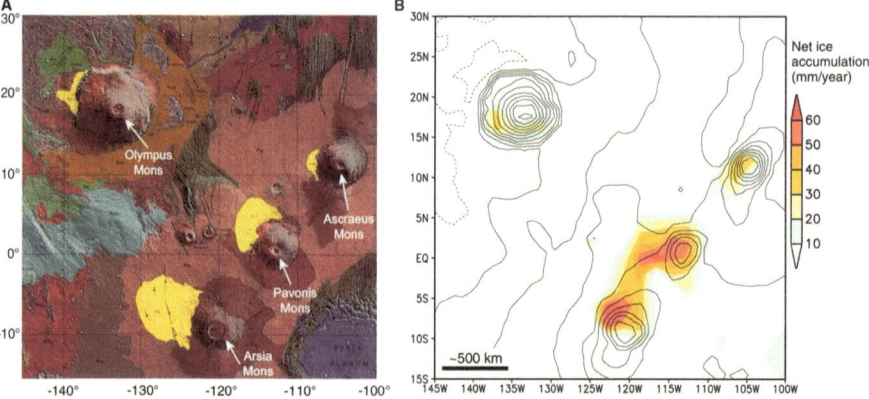

Fig. 3.13 On the left, a geological map of Mars showing in yellow the formation associated with ancient glaciers. On the right, accumulation of ice according to a climate computer simulation. This is a spectacular application of a climate model of Mars. In the simulation, snowfall accumulates on the windward side of the giant volcanoes on the Tharsis plains. And indeed on real Mars, rock formations indicative of former ice deposits have been observed in the expected locations. Image credit: A D.H. Scott, K.L. Tanaka, USGS. B Lab. de Meteorologie Dynamic, IPSL (Paris, France) MGS MOL

These climate simulations work by dividing the atmosphere into thousands of small cells. Physical quantities like pressure, temperature, composition, and cloud content of each cell are chosen at the start, and the laws of physics governing the interaction, motion and evolution of the cells are coded into the programme. The computer programme is then left to run forwards by small steps and scientists can see how the system evolves.

This does not always work out well, because so many physical processes are involved in an atmosphere. Not only are the processes numerous, but they occur over a vast range of distance scales, from microscopic turbulences to planet-wide storms, while time scales vary from seconds to years.

Furthermore, as we shall see later, the flow of gas in planetary atmospheres is turbulent. Turbulent motion involves a fundamental unpredictability, even with good knowledge of the initial conditions.

Martian dreams

A large part of the fascination with Martian landscapes comes from the way they echo landscapes on Earth, only on grander scales. Mars' Grand Canyon is vastly larger than its Colorado namesake. Mars' volcanoes are three times higher than Mount Everest. Mars also shows traces of gigantic flooding, possible relics from earlier times when the atmosphere was thicker and the conditions more Earth-like.

Vistas on Mars are also more Earth-like than in any other place in the Solar System, such as the view from the Viking landers, with a touch of morning frost on the ground, a red sunrise, and a couple of haughty cirrus clouds high in the sky.

One tenacious idea not confined solely to science fiction, but also found in other circles, is the idea that Mars could be transformed into a habitable planet with a thicker and warmer atmosphere, "terraformed". This stems from the realisation that, if only one could magically restore a thick atmosphere in Mars containing enough water and carbon dioxide, the greenhouse effect would be powerful enough to bring the temperature above freezing in spite of the far greater distance from the Sun. Lakes could then form, and seas, colonies, roads, cities and so on. Mars would be the new frontier where the epic history of the New World could be played out all over again, albeit without indigenous tribes, buffaloes and fabled cities of gold.

Fig. 3.14 Morning frost on Mars seen from the Viking lander. Image credit: NASA

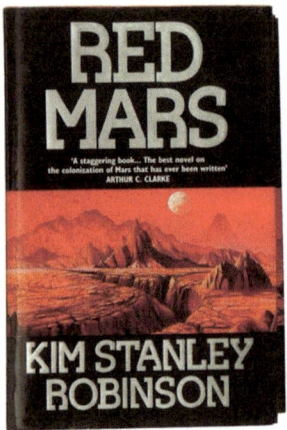 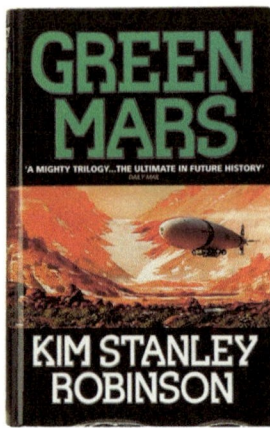 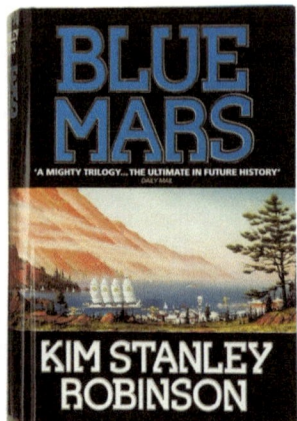

Fig. 3.15 A famous fictional example of the terraformation of Mars.

In the *Red Mars, Green Mars, Blue Mars* cycle, colonists use plants and melt the underground permafrost to terraform the planet. In the movie *Total Recall* based on Philip K. Dick's *We can remember it for you wholesale*, subterranean glaciers are simply heated until they reform an atmosphere.

In both cases, the timescales of physical processes were compressed by poetic licence, from thousands or millions of years to decades in the first case, and … minutes in the second.

The more one learns about the real Mars, however, the more this dream fades into the distance. Since Mars has one percent of the atmosphere on Earth, terraforming implies adding another 99 percent of atmosphere. This is not a very different proposition to that of adding an entire atmosphere to an airless planet. Mars is a frozen, airless, frigid planet, a cratered pile of rubble. The thin veil of CO_2 apart, it resembles the Moon more than the Earth. And at least there is a lot more Sun on the Moon[7].

The dream of terraforming Mars to create an outlet for human overpopulation seems to be based not only on shaky physics, but on dodgy geography as well. As inhabitants of places like Libya, central Australia, northern Chile or Siberia know well, what people need is not more room, but liquid water and a mild climate. There is plenty of room left on Earth, and nobody is lobbying too hard for the colonisation of the Franz-Joseph archipelago in the Arctic Ocean or the interior of Namibia, where room is plentiful. The fact is that it would be much easier – and cheaper – to send millions of people to live at the bottom of the sea or in the middle of Antarctica, than on Mars. Compared to Mars, the Antarctica ice fields are a virtual paradise: water in abundance, temperature generally above –50 degrees Celsius, and fully oxygenised air.

But like a lover who idealises the object of his/her dreams and turns even the most glaring shortcomings into qualities, we forgive Mars everything, and the more we study it, the more we find it welcoming.

---

[7] Before we think of terraforming the Moon, it is fair to say that the gravity of the Moon is too small to keep an atmosphere.

Fig. 3.16 Shalbatana Vallis of Xhantis and the south of Chryse Planitia from the Mars Observer Laser Altimeter, on the same scale as the next map; altitude-coded (with highlands in green and lowlands in blue). Image credit: NASA

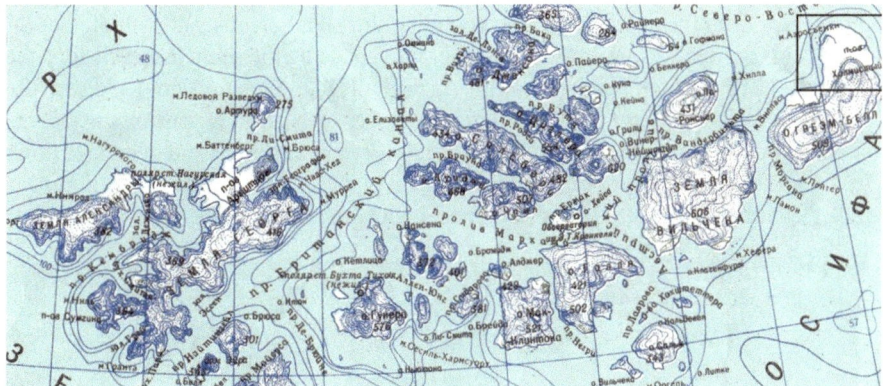

Fig. 3.17 Franz-Joseph Land, 400 kilometres (250 miles) across, abundant water, room for dozens of domed cities. Population zero.

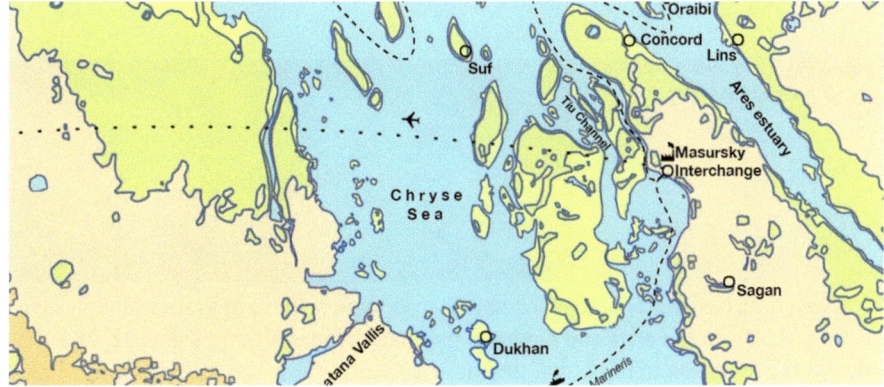

Fig. 3.18 The Chryse region after terraforming, showing the Masursky Interchange and the ferry terminal to Valles Marineris.

Three siblings

Most of the features of the atmosphere of Venus, Earth and Mars can be understood through the operation of physical processes over billions of years, Venus losing its water to space, Earth capturing carbon dioxide and sulphur inside rocks and oceans, and Mars losing almost its entire atmosphere because of low gravity. When scientists try to reverse this evolution and imagine what the three planets might have been like when they were much younger, their atmospheres seem to converge. It is then a small step to imagining that the three planets could have started in similar conditions at the birth of the Solar System, around 4.6 billion years ago.

The initial atmospheres would be thick and mainly composed of water and carbon dioxide. Vast seas would have covered the lowest part of the surface, as volcanoes spewed out more water and CO_2 into the air.

Several aspects of young Venus, Earth and Mars are still unknown. A few decades ago it was thought that the primitive atmospheres of the terrestrial planets were dominated by hydrogen, left over from the initial accumulation of material forming the planets from the nebula of gas surrounding the Sun. Scientists now think that this initial atmosphere was entirely thrown out into space by the immense collisions that mark the last stages of planet formation. As the balls of molten rock cooled down after the last impacts, gases that can dissolve in hot lava, such as water, carbon dioxide and sulphur, would have slowly vented out through volcanic eruptions and giant lava flows. In the end the atmosphere of young terrestrial planets (a few hundred million years old) would be made up of volcanic fumes, mainly carbon dioxide and water vapour.

Fig. 3.19 Venus, Earth and Mars in their early life (maybe) …

Fig. 3.20 … and after 4.5 billion years. Image credit: NASA

We might easily accept the fact that the toxic atmospheres of Mars and Venus were not present from the start and emerged gradually from volcanic exhausts, but it is a shocking realisation to think that the same is true of Earth. Our refreshing blue air is issued from the smelly, toxic smoke of volcanoes, re-processed by water and by life.

Chapter 4
Titan

Titan is the second largest moon in the Solar System, after Ganymede, which is Jupiter's largest moon. With a diameter of 5,150 kilometres (3,200 miles), compared to 3,470 kilometres (2,156 miles) for our Moon, Titan is even larger than the planet Mercury. Uniquely amongst moons, Titan possesses an atmosphere. From the outside, it looks like a yellow-orange, featureless globe, because its thick, hazy atmosphere entirely conceals its surface.

Fig. 4.1 Titan. Image credit: NASA

Fig. 4.2 The Sun seen from Titan (right), compared to its size from Earth, Venus and Mars (from left).

Seen from Titan – at almost ten times the Sun-Earth distance – the Sun is only a weak spotlight in the sky, and the temperature is correspondingly colder, around two hundred degrees Celsius below zero.

The main component of Titan's atmosphere is nitrogen, the same gas that dominates our atmosphere. The rest is methane (CH_4), hydrogen (H_2), and ethane (C_2H_6), another component of natural gas, with methane. At these low temperatures, heavier molecules like water and carbon dioxide freeze and form blocks of ice on the ground. The whole planet's crust is actually dominated by water ice, with maybe a buried liquid ocean, deep underneath.

The atmospheres of Earth and Titan are similar in several ways. They are the only two known atmospheres which are dominated by nitrogen. The pressure at the surface is slightly higher than here, at 1.54 atmospheres, a pressure we could adapt to (equal to the pressure under about six metres of water). The main difference is that Titan is so much colder.

Surface features

On 14 January 2005, the small Huygens probe pierced the haze of Titan and plummeted towards the surface. While parachutes were starting to slow down its fall, it took the first snapshots of the surface of Titan, never seen before by human eyes. The results were breath-taking.

The first images from Titan were more Earth-like than anything ever seen before. Titan may be a frozen, sterile world, but from the air it looked strangely familiar.

What did Huygens see? Rivers, streams, quiet shaded valleys, light morning fog. In the suggestive shadows of the first images one could almost imagine sloping forests, scattered juniper bushes or a fishing village.

The rivers are not made of water. Water is always rock hard on Titan because of the extremely low temperatures. What is flowing there, even forming lakes (that most

Earth-like feature) is liquid methane. Indeed, Titan is the only known place with liquid on its surface apart from our planet.

Fig. 4.3 Titan's "shoreline", visible on an early image from the Huygens descent. Image credit: NASA

During its fall, the Huygens probe swayed wildly because of its parachute; as a result its camera covered a good fraction of the landscape, if in a rather haphazard manner. Once on the ground the probe took a more detailed picture of its surroundings, showing it had landed in a somewhat disappointing spot – the usual Atacama-desert-style plain of rubble, not unlike Mars, the Moon, or Venus. No methane waterfalls in sight; no tungsten tree or cryogenic sheep.

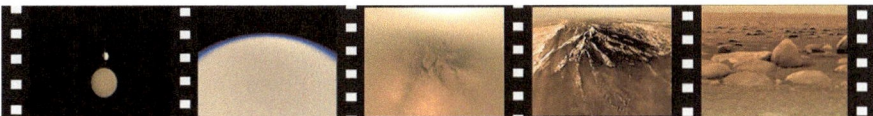

Fig. 4.4 Stills from the movie of the Huygens descent. Image credit: NASA

The planetary explorers

The engineers of the Huygens project had an interesting challenge to face: the presence of lakes or seas of liquid methane were expected on Titan, and they had to devise a probe that could either land on solid ice or splash into a sea of methane. Building a lander that can travel across the Solar System and parachute into an alien atmosphere is hard enough, but this was the first time the vehicle had to be amphibious.

The Huygens probe was equipped to float on an ocean of liquid methane. It would have frozen fast in such a situation, lasting no more than a few minutes. In the event, it landed on hard ground and was able to send data for about 90 minutes.

Upto 2012, some 14 vehicles have managed to print their little podmarks under alien skies: five on Venus, seven on Mars, and one on Titan.

The main obstacle for a probe to arrive alive and well on another planet is the speed of its fall. Different atmospheres pose different problems. On Mars the thin air does not provide much breaking and large parachutes and retro-rockets are used to slow down the fall. In some cases an airbag system was used to cushion the impact of contact with the ground, which can still be rather violent. In the thick, viscous air of Venus, slowing down the fall is not a problem, and the probe can gently sink below the clouds under a small parachute. The problem there is posed by the intense pressure and temperature, which pressure-cook the instruments in a few minutes. The central part of the Venera probes is a sphere of metal, similar to submersibles like the Bathyscaphe that reached the deepest parts of the oceans on Earth.

On Jupiter, in 1995 the Galileo probe fell through an increasingly thick atmosphere with no solid ground in sight, until it entered gas layers so dense that contact with Earth was lost. Since its metal components remained denser than the surrounding hydrogen at any pressure, the probe will have continued sinking, ever more slowly as the density increased, until the components were melted by the high temperature and crushed by the high pressure, and carried around in the planet's huge convective motions. Over time, the scattered pieces of metal will have reached regions deep inside the planet where the temperature is high enough to melt and vaporise them and combine their individual atoms in the general mix of Jupiter's composition.

On each planet, the fate of the landers gives an indication of the atmosphere. Vulnerable parts of the Venera probes softened and buckled under the stifling Venusian conditions, while the metallic parts were slowly corroded by the acidic atmosphere until the whole craft was reduced to lumps of metallic oxides blending with the local rock landscape, its exotic composition the only reminder of its alien nature. The Martian landers on the other hand will remain in shape for millennia, like the vestiges of the Pharaohs preserved in the arid Egyptian desert, the oxidising air and tiny dust particles very slowly chipping their metal plates away. They may end up covered by their own little sand dunes before disintegrating.

Contrast these with the pieces of equipment that the Apollo astronauts left on the Moon, which are expected to remain in pristine condition for millions of years. These may well be the last human artefacts left in the Solar System long after the disappearance of any trace of human occupation on Earth.

Fig. 4.5 Titan methane lakes, reconstructed from measurements of the smoothness of the surface. Image credit: NASA

Lakes of methane

Instruments aboard the Cassini spacecraft (which launched the Huygens probe in 2005 and has aproached Titan several times since then) have revealed how smooth or rough the surface is on various parts of the planet. It is thought that especially smooth patches are lakes of methane.

Close observation of one of these lakes, Ontario Lacus, has even shown the shoreline of the lake receding with time, a sign of the passing seasons on Titan.

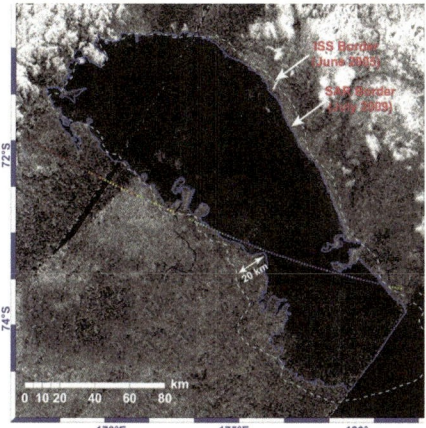

Fig. 4.6 Receding shoreline of the methane lake Ontario Lacus on Titan. Image credit: NASA

Methane cycle

There are clouds, rivers and lakes on Titan for the same reason as on Earth – because a substance changes between its solid, liquid and gas forms with changes in temperature. On Earth this substance is water, on Titan it is methane. Methane melts at –182 degrees Celsius and boils at –162 degrees Celsius (compared with water's 0 and 100 degrees Celsius).

Such cycles are often what make a planetary atmosphere interesting. We have encountered the water and carbon cycles on Earth, the CO_2 cycle on Mars and the two sulphur cycles on Venus. Carbon dioxide on Mars moves from its ice form in the polar caps, to being a gas in atmosphere, where it form clouds, then snows down in winter and regulates seasonal temperature changes. Sulphur on Venus forms droplets of sulphuric acid that shroud the whole planet in thick clouds.

On Earth, a large fraction of the energy from sunlight is used to evaporate water from oceans, lakes and forests rather than being turned into heat, so that the water cycle acts as a giant temperature regulator. In the same way that sweating is an efficient way of cooling a human body, evaporation prevents oceans and landmasses from overheating. Running water is the major factor in land erosion, shaping continents and returning sediment to the sea. Less visibly for us, the water cycle also plays a key role in regulating the amount of carbon dioxide and the warming in the atmosphere through the greenhouse effect.

Methane lasts only one or two decades in the atmosphere of Titan, after which it

Fig. 4.7 Profile of the atmosphere of Titan, on the same pressure scale as Earth.

is modified by interacting with the rocks or the sunlight, forming for instance acetylene (H_2C_2), ethane (C_2H_6) or more complex molecules like polymers or aromatics. It must therefore be constantly replenished by a cycle. Compared to the Earth's cycles, the water cycle with timescales of hours, and the rock-carbon cycle with timescales of millennia, the methane cycle occupies an intermediate position.

The driving force of the water cycle on Earth is the evaporation of vapour from the surface of the ocean, occurring most intensely over the few hottest hours in the equatorial region. The driving force of the rock-carbon cycle is the volcanic production of CO_2, a slow and episodic process related to the secular motion of the tectonic plates and the convection of the mantle of the Earth over millions of years. The driving process for the methane cycle of Titan is the chemical reaction of the methane molecules with other substances in the atmosphere or on the ground, or with sunlight above the haze.

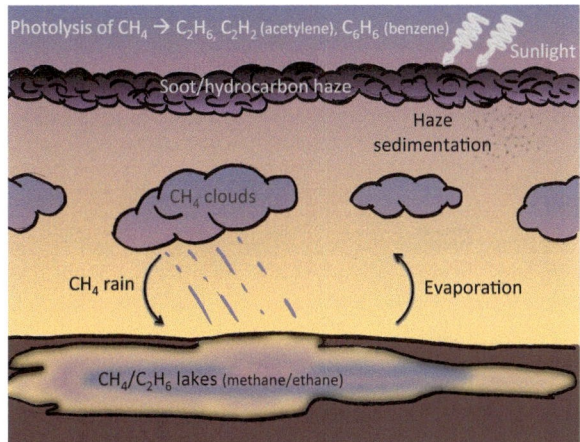

Fig. 4.8 The role of methane in the atmosphere of Titan. Image credit: Joanna Barstow

Iceball Titan

Methane is essential to Titan's atmosphere in another respect. It is a powerful greenhouse gas, transparent in visible light but very opaque towards infrared light. Even the tiny amount of methane that we have in Earth's atmosphere makes an important contribution to the greenhouse effect here, so with methane being the most abundant component after nitrogen on Titan, the greenhouse effect is large. Although the temperature is only –180 degrees Celsius, it would be much lower without methane. So low, in fact, that the temperature would dip below the condensation temperature of nitrogen (–196 degrees Celsius), leading to the collapse of the whole atmosphere. Liquid nitrogen would start flowing and freezing out of the sky, until the whole planet became a frozen, airless body like the other satellites of Jupiter and Saturn.

Maybe one day the methane content of Titan's air will drop below a critical value that will lead to the collapse of its atmosphere. Would Titan then be able to kick itself out of its frozen state and raise the temperature enough for the nitrogen to evaporate and re-form the atmosphere? This is possible, because volcanism keeps injecting methane and other gases into the atmosphere.

In fact this tells us that higher gravity is not the only thing that enables Titan to maintain a thick atmosphere while other slightly smaller satellites of Saturn and Jupiter cannot. Its larger size means that it is also able to keep more internal heat, and therefore maintain more volcanoes to regularly eject gases and replenish its atmosphere.

Cryovolcanism

The volcanoes on Titan are rather different from what we have on Earth. What is spewing out of the frozen ground is not lava and sulphuric gases, but water vapour, methane, ammonia and nitrogen. Because of the extremely low temperatures, in the outer parts of the Solar System, ice plays the role of rocks. Water is a very solid material at –200 degrees Celsius and only far hotter temperatures rising from deep inside the planet can make it evaporate, and explosively escape towards the surface. This is aptly termed *cryovolcanism*.

Fig. 4.9 Enceladus "geyser"
spraying water vapour directly
into space. Image credit: NASA

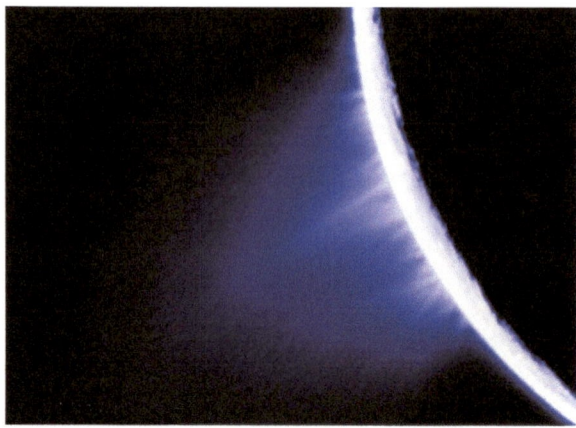

Fig. 4.9 Enceladus "geyser" spraying water vapour directly into space. Image credit: NASA

   Cryovolcanism is common on the satellites of Jupiter and Saturn. Enceladus, one of Saturn's smaller moons, regularly emits jets of water vapour and other cold gases into space. In this case the gravity is too low to keep the volcanic plume on the moon, and the gases from volcanoes disperse in space to form a trail of gas around Saturn.

   Hot volcanism affects the inner circle of planets (Mars, Earth, Venus and Mercury), and cryovolcanism planets of the outer solar system, but there is one exception: Io, the first satellite of Jupiter, has a rocky surface and hot volcanoes. The reason is that Io is so close to Jupiter that its interior is churned up by its tidal pull to such an extent that it is heated from inside and has become warmer than Titan; so warm in fact that ice has entirely disappeared from the planet. Earth-type volcanoes regularly spit out sulphur, CO_2 and water vapour. Io's gravity is too weak for it to be able to cling onto its gases, so it has no atmosphere and most of the volcanic smoke drifts straight out into space. Only the sulphur compounds, being heavier, fall back onto the surface.

Fig. 4.10 A volcanic erruption on Io. Image credit: NASA

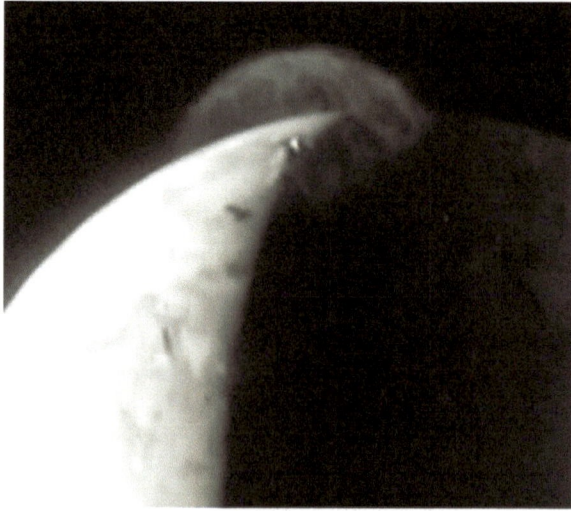

Titan's haze

From space, Titan looks like a smooth orange sphere with fuzzy edges, because of the thick haze that shrouds the whole planet. The haze on Titan is mainly composed of the product of the photochemistry of methane and ethane. Several of these carbon-based molecules have been detected on Titan by the Huygens probe. Strangely enough, many of these are constituents of oil and gas fields on Earth: propane, acetylene, benzene, etc... Titan's weather is a constant "oily drizzle".

If Venus is eerily reminiscent of the medieval image of hell, and Mars is a frozen arctic wasteland (the hell of some Nordic mythologies), then Titan is a more modern kind of hell, a super-cold industrial nightmare.

Photochemical haze vs condensation clouds

Clouds form when part of the atmosphere rises up and cools below the condensation temperature of one of its components. Water, sulphuric acid or ammonia forms grains or droplets when condensing. Because of the requirement of rising currents and temperature changes, clouds are usually associated with atmospheric motions, hence their wonderfully varied shapes and structures.

The haze above Titan is not made up of droplets that condense because of the low temperature, it forms following chemical reactions triggered by the light of the Sun. Chemical reactions that use sunlight are a key element of atmospheric physics that is very remote from our daily experience. They are instances of what is called *photochemistry*, photo being Greek for light.

In ordinary chemistry, as when cooking a stew, boiling an egg, or making aspirin in industrial tanks, some molecules can interact with others to swap atoms and electrons and turn into other molecules. These reactions are generally controlled by temperature. Temperature – random motions of atoms and molecules – is needed so that molecules bump into each other with sufficient velocity to overcome some of the initial repulsion of electrons, until the outer electrons can feel the positive charge of the protons in the nucleus of the other atoms. That's why many reactions have to be started with the heat of a flame, like the combustion of natural gas or wood, until they produce their own heat to keep it going. At absolute zero or –273 degrees Celsius (the temperature at which all atoms stand still) no chemical reaction can happen.

Heat is not the only way to bring enough energy to initiate a chemical reaction, however, sometimes the energy can come directly in the form of light. In photochemistry, an atom or molecule absorbs light to trigger a chemical change.

This happens in photosynthesis, which allows plants to draw energy directly from sunlight. One particle of light, a photon, gets captured by the chlorophyll molecule. It modifies its structure so that it will be able to transmit energy through electron-swapping to a long chain of molecules, to provide useable power for the plant.

The retina in our eye is also able to capture light directly, in reactions that ultimately produce electric currents in nerve cells to inform the brain about the layout of our surroundings.

We have already encountered two instances of photochemistry in planetary atmosphere. The ozone layer on Earth is caused by the reaction of oxygen with sunlight.

On Venus, sunlight dissociates the sulphur dioxide in the atmosphere into more active molecules, which then react to form the grains and droplets that make up the clouds of the planet.

At ground level on Earth, photochemistry can act on some chemical pollutants to form larger compounds, which can aggregate into particles and give a perceptible brownish hue to the air – the industrial smog that can form over cities on windless summer days.

On Titan, the photochemical haze contains dozens of carbon compounds. Why don't these complex carbon molecules form in Earth's atmosphere? After all, there is much more sunlight on Earth, and at least as much carbon.

The key difference is a spectacular feature of our atmosphere that we have encountered in Chapter 1: with 20 percent oxygen, our atmosphere is so reactive that complex carbon compounds cannot last long in the air. Their carbon chains get ripped apart and carbon atoms are captured by oxygen to form CO_2. In everyday language, they burn. Our atmosphere is so thick with reactive oxygen that anything that can react with oxygen will do so, given time.

Life in liquid methane

With such a rich, carbon-based chemistry and the most Earth-like weather system we know of, could Titan possibly harbour life?

Not our kind of life; life on Earth requires liquid water, and on Titan all the water is frozen. Water is essential to life as we know it, as a solvent and a matrix in which all the molecules of life can interact and go about their business. A liquid solvent is thought to be essential to the chemistry of life, because chemical reactions in solid material tend to be slow and difficult, while in gases they are too random and scattered. Water provides the gentle but firm continuum that proteins, nucleic acids, sugars and lipids require to prosper.

Could liquid methane do the trick? This is a very difficult question, since nobody knows how a methane-based life system would work. It is difficult enough to understand life on Earth, and with only one example of life having appeared, it is even harder to distinguish the essential from the accidental. Nevertheless, using our present understanding, some people have tried.

As a solvent, methane is weaker than water. Because of its lower temperature and weaker electric properties, methane can dissolve smaller molecules than water can. The kinds of sugar that we use as a source of energy, for instance, would be too large for methane, and would simply sink. There are, however, a good number of carbon compounds that can dissolve in methane, including some phosphate molecules that would be too fragile at our kind of temperatures.

How many molecules would be enough to sustain life? One attempt to address this question is to consider how many different molecules are used in the basic operations of a living cell on Earth. Present estimates put this number around 700. Many more carbon compounds dissolve in water, so life on Earth has a comfortable selection of molecules to choose from. In methane the total number of molecules that can dissolve is only a few hundred, which may be sufficient, but barely.

At present I think most specialists would say that methane-based life is not entirely excluded, but requires stretching the chemical possibilities to their limits.

How would we detect methane-based life on Titan? Short of spotting a methanoid moose crossing the field of vision of a probe's camera, the best way is to analyse the composition of the atmosphere and look for the kind of disequilibrium that life has created on Earth. In the same way that life on Earth has given itself away by poisoning the atmosphere with 20 percent of reactive oxygen, the methane life of Titan could be producing unstable compounds which we could detect in its atmosphere.

The big question when pondering the possibility of life outside Earth is to assess how earth-centered we are in our assumptions about life in general.

Living atmospheres

There is probably no life on Titan. Nevertheless, there is another way, more metaphorical, in which it can be helpful to consider some atmospheres as a living system.

Sometimes, when some systems become complex enough, they start behaving like more than the sum of their parts. Think of an *ecosystem* on Earth for instance. It may be made up of living creatures, but the ecosystem as a whole is not a living creature. Scientists have found that complex ecosystems can be very resilient. The intricate relationships between their constituents makes then able to deal with circumstances in a much better way than simple systems.

An important feature of the properties of complex systems, that makes them seem alive in a certain sense, is the presence of cycles. Feedback loops allow regulation. Cycles like the methane cycle or rock-carbon cycle introduce such loops into planetary atmospheres, and some researchers have suggested that the analogy with living systems may be useful.

At the extreme, the "Gaia" view of the Earth posits that the atmosphere of our planet has now reached such a state of intricacy that it functions more like a living system than a physical system. Feedback loops allow it to control its temperature and the composition of the atmosphere, maintaining the optimal conditions for life.

An early example of how this kind of system could work was given by the imaginary *Daisy world* planet. This is a simple model of a planet covered with plants. When the temperature gets too hot, some plants die. This makes the surface of the planet brighter, because deserts and rocks are lighter in colour than forests and vegetation. As a result, more sunlight gets reflected into space, and the temperature cools. In Daisy world, life regulates the temperature of its planet through a simple feedback mechanism.

Living organisms use this kind of regulating mechanism very often; there are many examples in our own body, control loops that allow us not only to maintain a constant temperature, but also fight off infections and keep track of time.

The atmospheres on Earth and on Titan seem complex enough for such system-level properties to be present.

Is this true for all atmospheres? Probably not. For instance, the atmosphere of Mars is so light that it doesn't seem to have the capacity to control its own fate. In summer, intense sunlight can send the whole planet into a dust storm. In winter, the atmosphere collapses onto the poles.

Whether complex planetary atmospheres – and that of Earth in particular – can be well understood as physical systems, or whether they can also be likened to living systems, have important consequences for us as we try to come to terms with the issue of climate change.

In the first case, climate models tell us that an increase in the carbon-dioxide content of the atmosphere will heat the climate catastrophically. In the second case, however, the Earth's climate is much more resilient than we think, and feedback mechanisms will react to keep the climate around its present state. This is not necessarily as good news as it sounds, because resilient systems tend to absorb strain up to a certain point, then "snap" abruptly (not unlike the way the human brain deals with trauma). In that case, our atmosphere may succeed in handling the changes for some time, but there may be a snap lurking just behind the corner.

At present, the balance of evidence is rather against the existence of a "Gaia" condition for Earth, at least from the point of view of global warming. The best evidence is provided by an event that happened at the dawn of the age of mammals, in the Eocene some 60 million years ago. This event and its consequences on the climate have been recorded in detail by rocks and sediments.

A huge volcanic eruption at that time briefly doubled the concentration of CO_2 in the atmosphere. Global temperatures then spiked by 4–5 degrees Celsius for a few thousand years, as measured by indicators such as the concentration of oxygen isotopes in sea shells. This is as expected if the climate was reacting passively to the added greenhouse effect, without any feedback mechanism to mitigate the change.

Another indication is the occurrence of planet-wide ice ages, or "snowball episodes", in the history of the Earth. In this case the feedback goes in the wrong direction: colder temperatures increase the ice cover, which makes the planet more reflective and reduces the amount of sunlight absorbed. It seems that life and the atmosphere were not able to do anything about it in the past. The planet was brought back to a milder climate only by volcanoes, not by any sophisticated resilience mechanism.

Being there

What would it be like to stand on Titan? The cold would be terrible, but we could shelter in sophisticated heated habitats like those of the Concordia research station in Antarctica. The air is mainly made of nitrogen like on Earth, which is nice, and the atmospheric pressure is bearable. Neither methane nor ethane are toxic, and they are not flammable on Titan because of the lack of oxygen. We would need oxygen to breathe of course, but light oxygen masks like in hospitals or airplanes could do the trick.

Outside everything would be rather dark, because the Sun is so far away. The sky would be a pale, uniform glow. Days last for 15 Earth days, so there would be one week of darkness every fortnight.

One detail that we may notice outside is that sound travels much more slowly in very cold air. The horn of the returning exploration vehicle would sound much lower than normal (the opposite of the "funny voice" effect of Helium gas), and the noise of construction machines in the distance would seem to take forever to reach us.

Chapter 5
Giant planets

Four giant planets dominate the Solar System family. Their colossal scale is difficult to grasp. More than one thousand Earths would fit inside Jupiter. The giant planets represent 99.5 percent of the mass of planets in the Solar System, with Jupiter totalling 318 times the mass of the Earth, Saturn 95, Uranus 14 and Neptune 17. Their truly majestic nature is one of the most distinguished discoveries of space exploration in the second part of the twentieth century.

Fig. 5.1 Jupiter, Saturn, Uranus, Neptune, Earth, Venus, Mars, Mercury and the Moon. Image credit: NASA

This majesty is reflected in their names, and the echoes of Greek myths. Jupiter, the Roman name for Zeus, is the King of the Gods and god of thunder and lightning (he is often depicted wielding thunderbolts). In the Iliad, when faced with a mutiny by the other Olympian gods, Zeus challenges them all to hang on one end of a rope, while he alone pulls on the other end. He boasts that he could lift them all up, as well as all the land and sea, then "leave them dangling in space".

This is a fitting metaphor; the planet Jupiter is heavier than all the other planets combined, and could scatter all of them out of orbit if they came too close (some scientists think this actually happened, see page 105). A wiggle in Jupiter's orbit could scatter the lesser planets and send Mercury, Venus, Earth and Mars hurtling through the Solar System like mere comets.

Saturn was Jupiter's father and was known as Kronos in Greek. Warned by a prophesy that his sons would overrule him, Kronos devoured his children as soon as they were born. But when his wife Rhea gave birth to Zeus, she managed to trick

Kronos into swallowing a stone instead. In due course, Zeus grew up to lead a revolt of the gods against his father.

As for Uranus, he was Kronos's father and the god of the sky itself. So the story goes that Uranus and Gaia were the parents of Kronos (and many others), who in turn fathered Zeus (among several others), each son replacing his own father after an episode of violence.

The actual planets stand up to these magnificent tales.

Gas and ice giants

Jupiter and Saturn are known as *gas giants*. They are made mostly of a mixture of hydrogen and helium, the material that stars are made of too, and have no real surface. The name *gas giant* can be misleading though, because these planets are not fluffy balls of gas. In their interior, the pressure is so extreme that the hydrogen/helium mix is compressed to very high density, and starts behaving more like a metal than a gas.

Neptune and Uranus, by contrast, are known as *ice giants*. From the outside they look similar to Jupiter and Saturn, with an atmosphere of hydrogen and helium, but their interiors consist mostly of dense water, methane and ammonia, or "ices" in the jargon of planetary science.

The formation of planets

Planets form by aggregation of dust and gas in a swirling disc around a nascent star. Stars like the Sun are always surrounded by a disc of gas and dust as they form. Close to the star, only rocks and metals condense. They gather into particles, then grains, pebbles, asteroids, and finally, through random collisions, an Earth-like planet can form. Further away from the star, the temperature is low enough for lighter substances like water, methane and ammonia to condense. These compounds are much more abundant than metals and rocks in the interplanetary gas, therefore far larger planets can form, such as Uranus and Neptune, which are mostly made of water. When a solid planet becomes heavier – a few times heavier than the mass of the Earth – the nascent planet is heavy enough to capture the remaining gas in the surrounding disc and swallow everything in its reach. This is what Jupiter and Saturn have done.

At some point the nascent star becomes bright enough to blow the disc of gas away and halt the growth of its planets. The whole process takes a few million years.

Because of these diverse formation processes, atmospheres can have different origins. The gas envelope of giant planets is made up of material captured directly from the disc of gas around the new-born star. Terrestrial planets, by contrast, formed by accumulation of asteroid-like elements, remain too small to retain the gases they had at birth. Any atmosphere they might have had would be blown away into space by the impact of their formation. Any subsequent atmosphere comes from the hot interior, through successive volcanic eruptions. Some of it may also come from later, gentler impacts of other asteroids and comets wandering through the early planetary system. For instance, it is thought that some of the water in the Earth's oceans may have come from comets, which are known to be mainly made of water in ice form.

Let us take a look again at the "Astronomer's Periodic Table", with the elements arranged according to their abundance in interstellar gas. This is the basic material which is available to form planets around most stars. According to the table, the most abundant atoms here are hydrogen, helium, oxygen, carbon and nitrogen. These atoms will tend to combine as water (H_2O), methane (CH_4), and ammonia (NH_3), with the noble gas helium staying aloof.

Fig. 5.2 (1.12 repeated): Astronomer's periodic table.

The temperature in a disc of gas at 150 million kilometres from the Sun (the same distance as between Earth and the Sun, i.e. one Astronomical Unit) is around 0 degrees Celsius, and hydrogen, helium, water[8], methane and ammonia are in vapour form. The main solids available are compounds containing silicon, magnesium, aluminium, iron, calcium or sodium. Most of these metals condense as oxides, forming what we would generally recognise as "rocks". Some heavier metals like iron and nickel can also condense in metallic form.

The asteroids roaming between the orbit of Earth and Mars since the birth of the Solar System are made of a mixture of rocks and pure metal. This is also, in fact, what the Earth itself is made of – a core of liquid iron and a mantle of rocks, mostly silicates. Although it is dear to us, the thin layer of water and gas on the outside represents less than a tenth of a percent of the total mass of the planet.

Further away from the Sun (3-5 Astronomical Units and beyond), the temperature becomes low enough for water, methane and ammonia to condense. These familiar molecules condense into solid ices, and therefore become available for the formation of grains, then bigger lumps and clumps, and finally planets. In interstellar gas, oxygen, carbon and nitrogen are far more abundant than silicon or metals, so they will dominate the composition of grains in the cold regions of the immense disc, which stretches out from the surface of the star to hundreds of Astronomical Units. The comets scattered in the far reaches of the Solar System are mainly made of water ice, with some frozen methane and ammonia, mixed with a few pieces of rocks or iron in the centre. Except for the fact that they are a lot smaller, they are similar to the frozen moons of Jupiter and Saturn.

Gas capture and ice line

A planet can only attract and keep hold of an atmosphere if it is heavy enough for its gravity to keep the gas from floating back into space.

Planets the size of Earth can keep hold of nitrogen, oxygen and carbon dioxide, but not of lighter elements like hydrogen and helium. Since most of the gas in the disc

---

[8] Water freezes at zero degrees on Earth, but in the vacuum of space it freezes only below –60 degrees Celsius.

around a nascent star is hydrogen and helium, the Earth did not capture very much of these gases.

It requires around ten times the mass of the Earth to be able to retain the lightest gases. When a young planet reaches this mass, it no longer needs to grow by accumulation of solids and collision of asteroids, but starts to swallow gas directly from the disc around it. This is a runaway process: the larger it becomes, the more its gravity is reinforced, and the more gas it can capture from the disc. The process only ends when a gas-free gap opens in the disc: the planet has captured all the gas it could acquire.

Saturn and Jupiter reached their enormous sizes in this way. At their core sits a mass of ice and rock around ten times larger than the whole Earth, surrounded by an envelope of 90 Earth masses (for Saturn) and 300 Earth masses (for Jupiter) of hydrogen and helium gas captured from the disc[9].

There were not enough rocks and iron in the disc around the Sun to form solid cores heavy enough to keep hold of hydrogen and helium. This is why giant planets did not form in the inner parts of the Solar System, where the temperature is too high for ice to condense. The outer parts, however, are cold enough for water, ammonia and methane to freeze and form solid bodies, and since these compounds are much more abundant than rock or iron in the cosmos, they could form cores large enough to reach the critical size which triggers the formation of a gas giant planet.

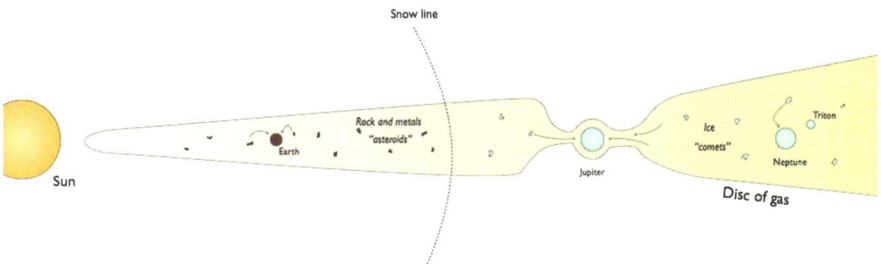

Fig. 5.3 Formation of planets and the snow line. Planets form by the accumulation of solidified material in the disc of gas that surrounds nascent stars. Close to the star, only rocks and metals can condensate. Further out, beyond the *snow line*, water condenses as ice. Larger planets can form. The largest bodies become heavy enough to accrete the gas from the disc directly.

The imaginary line separating the inner part of the Solar System – too hot for the condensation of water in space – and the outer reaches, is called the *snow line* (or *ice line*). Within the snow line, only rock-and-iron planets are expected to form, like the terrestrial planets Mercury, Venus, Earth and Mars in the Solar System. Beyond the snow line, gas giant and ice giant planets can form, as well as icy satellites such as Ganymede, Europa and Titan.

When small bodies visit the Earth hailing from other parts of the Solar System, we call them asteroids if they are mainly rocks and metals – these were presumably formed within the snow line. If they are mainly made of ice we call them comets – they presumably formed in the outer parts of the Solar System. The tails of comets are produced by the evaporation of ices under the heat of the Sun, as they travel towards

---

[9] For the record, it is not known whether Jupiter has a central core of rocks and ices, or whether these heavy elements are mixed up with the concentrated gases which form the bulk of the planet. We should know this in a couple of years, thanks to the JUNO mission which is on its way to the Jupiter system as this book goes to press.

warmer climes. But nothing evaporates from asteroids, since they were formed at relatively warm temperatures and do not contain ice or other material that could be easily vaporised.

Three types of planets

In broad terms, planets can form from the accumulation of asteroids, the accumulation of comets, or by the capture of hydrogen and helium gas by an already large core.

This is therefore what we would expect to find around a star with a disc rich enough in metal and ice grains to form planets:

> i. Small rock-and-iron planets in the vicinity of the star, closer than the vaporising point of water, ammonia and methane in space. In our Solar System these are Mercury, Venus, Earth and Mars; all these planets are closer than three Astronomical Units from the Sun.

> ii. Larger planets made mainly of water, methane and ammonia, with some rocks and iron, further out from the star. Examples: Uranus and Neptune, as well as the satellites of the giant planets such as Ganymede, Europa and Titan.

> iii. Giant planets made mainly of hydrogen and helium gas, with a rock/iron/water core of around ten Earth masses. Examples: Jupiter and Saturn.

Once planets are large enough to melt their interior, the heavier components tend to flow towards the centre and the lightest to float up to the surface. That is why planets adopt a "Russian-doll" structure, with an iron/metal core in the centre, a rock layer, then a layer of ices, and a hydrogen and helium envelope. Not all planets have all of these layers, but they always come in the same order. A planet with, for instance, a core of ice and a mantle of iron would not be stable, and the iron would sink to the centre over time.

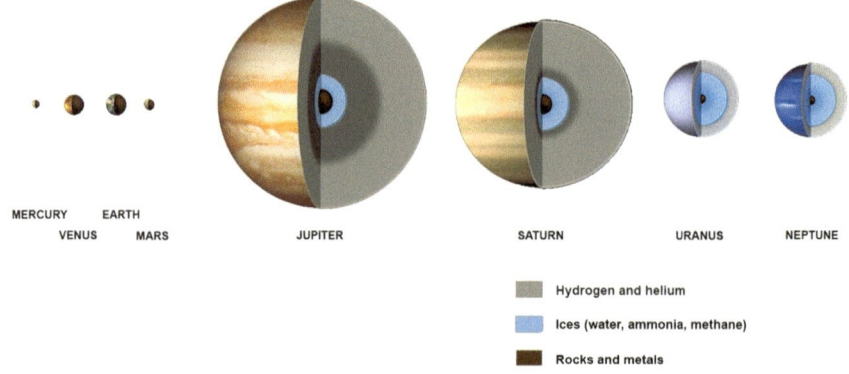

MERCURY   EARTH
    VENUS   MARS   JUPITER   SATURN   URANUS   NEPTUNE

■ Hydrogen and helium
■ Ices (water, ammonia, methane)
■ Rocks and metals

Fig. 5.4 Internal structure of planets in the Solar System.

Uranium-core planets?

If heavier elements sink more than lighter ones, what about the tiny fraction of very heavy elements, all the way down the periodic table, from gold and platinum to uranium? Shouldn't they sink to the centre as well? The central core of planets would then be composed of an onion-skin structure of pure heavy metals in ascending order of atomic weight. Are all planets powered by a natural nuclear reactor at their centre, a blob of pure uranium? In fact not.

Gravity is able to sort gases or liquids according to their relative weights, as can be seen in a vinegar-and-oil sauce. But atoms combine into molecules, and molecules stick to each other in solids and liquids, so that individual atoms are not free to move around as they wish. The links between atoms inside a molecule are stronger than gravitational stratification, so it is not elements that are sorted out by gravity, but molecules in gas, and associations of molecules in liquids and solids.

Heavy metals like uranium get trapped in molecules or alloys that tend to be less dense than iron. On Earth, they remain stuck as a very rare fraction of the lighter rocks in the mantle, and don't even make it into the iron-nickel metallic core. Uranium, like most metals, forms oxides, which aggregate as rocks, and these rocks are less dense than elemental iron. The Uranium of the Earth (0.000002 percent of the mass of the Earth, still about one hundred thousand billion tonnes) remains safely locked in rocks dispersed through the whole envelope of the planet.

Why is the inner core of Earth-like planets made of iron? This is due to a quirk of nuclear physics. Iron happens to be the most stable nucleus among all the elements. In the furnace of a supernova explosion – where most heavy elements in the cosmos are produced – the cataclysmic rearrangement of protons and neutrons favours the formation of iron over all other metals.

Deep inside

Hot concentrated ice, boiling iron, metallic hydrogen… the interiors of planets are made of familiar materials in unfamiliar states.

In our daily lives we are familiar with the way matter changes state under the influence of temperature. Ice melts, water boils, grease and honey turn solid in the fridge, and volcanic lava solidifies when cooling. We are less familiar with the way pressure affects matter, because pressure does not vary as much as temperature around us. We may have felt a little short of breath on the top of a mountain, or we may have experienced the pain of water pushing on our eardrums when diving, but these pressure changes are not large enough to affect the behaviour of common materials.

In the interior of planets, the pressure is enormous, millions of times that on the surface of the Earth, and materials become compressed into increasingly compact structures. Hydrogen is a gas near the surface, but in the depths of a gas giant the pressure is so great that it snatches electrons from individual atoms and forms a solid mass called *metallic hydrogen*. Scientists use this term because metals are characterised by the presence of free-ranging electrons in a lattice of atoms.

If you want to know what is inside a planet like Jupiter, you have to calculate how dense hydrogen would become under extreme pressure. Hydrogen is the simplest atom, consisting merely of one proton orbited by a single electron. Nevertheless, it turns out that the density of hydrogen at extreme pressures (millions of bars) is difficult to calculate, and has to be measured with experiments.

One of the ways to do this is to let a thermonuclear bomb explode, and measure the results. Indeed, some of the measurements on high-density hydrogen come from Russian thermonuclear bomb tests. Other, more gentle ways exist, such as using lasers to produce strong local shocks.

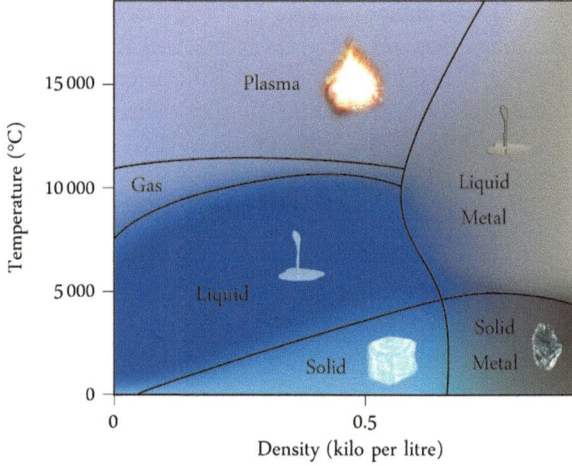

Fig. 5.5 States of hydrogen under different temperatures and pressures.

On ice giant planets such as Neptune and Uranus, the main constituent is water. The behaviour of water at high pressures is unexpectedly complex, with no less than twelve possible crystalline structures. We know one of them quite well, ice, which scientists call *ice Ih*, but this form of ice is just one among many. In giant planets, as the pressure increases, water is compressed into various semi-fluid, compact forms

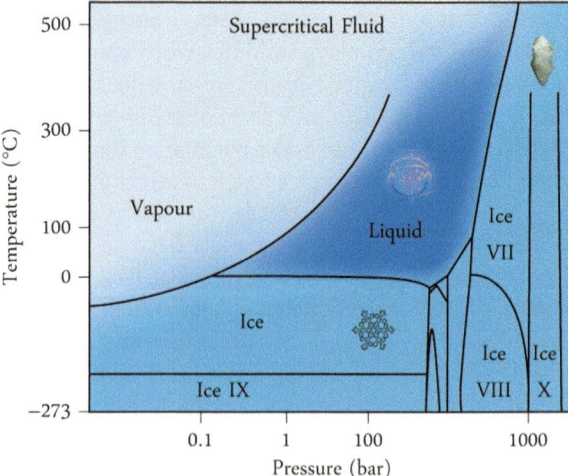

Fig. 5.6 States of water under different temperatures and pressures.

very different from ordinary ice. The interior of Neptune for example is thought to consist of a hot slush of water, ammonia and methane crystals, something heavy and sluggish like the magma found in Earth's mantle.

The interior of ice giant planets reaches several thousand degrees, as hot as the surface of the Sun; and lumps of hot ice from there, if ever brought to the surface, would emit a bright, blinding white light. A block of water straight from the mantle of Neptune would fry us in an instant.

There is another difference between ice giant planets and terrestrial planets: the lack of an identifiable surface. The material in giant planets transforms from a dense liquid to a thin gas without a well-defined phase transition. There is no boundary, the pressure and temperature gradually drop from the white-hot magma interior all the way to the thin gas of the atmosphere.

Atmosphere of Jupiter

We know the atmosphere of Jupiter quite well, thanks to the heroic sacrifice of the Galileo probe that plunged into the planet on 7 December 1995. Let us follow it in its descent.

When the probe first starts feeling the atmosphere, its thermal shield heating up due to the friction of the thin hydrogen gas, the air is clear and the temperature a chilly –120 degrees Celsius. Gradually, the probe sinks into thicker air and slows down. About 300 kilometres further down into the planet, the temperature starts rising, and we approach the highest tops of the first layer of giant clouds. These clouds are made of droplets of ammonia. A few dozen kilometres deeper, the probe crosses a second layer of clouds, made of a compound of ammonia and sulphur. Sunlight cannot penetrate the thick clouds and it is now pitch dark.

Finally, about 80 miles below the first cloud cover, the probe glides between giant clouds of water ("our" kind of clouds). The temperature has now reached a pleasant 20 degrees Celsius, but the pressure is crushing. The air is so dense that the probe no longer dives but sinks slowly like a pebble in a jar of honey, until it melts and dissolves in a thick, black, unsavoury brew.

Clouds

Clouds are the dominant feature in the atmosphere of giant planets.

We have seen that clouds form when the temperature drops below the condensation point of one of the components of the atmosphere, making it gather into droplets or grains. In a completely still atmosphere, these droplets and crystals simply drift down in a single episode of rain or snowfall, until the air is dry and clear again. That is why clouds do not only require the right temperature to form, but also atmospheric motions to bring the molecules that form the droplets back into the cooler regions.

Fig 5.7 Atmosphere profile for Jupiter, on the same pressure scale as the Earth.

We saw in Chapter 1 how this explained the rarity of clouds on Earth above the tropopause, the level at which commercial airplanes fly. The churning of the lower atmosphere by vertical convection and horizontal weather systems constantly moves new water from warmer to cooler regions, while this does not happen in the stable stratosphere.

On Earth, clouds almost never form in the stratosphere, although there are rare and spectacular exceptions. Sometimes over polar regions the air becomes so cold that even the low amount of vapour present in the stratosphere can condense. These clouds are called *iridescent clouds* because they are so thin that they only become visible in the grazing light of the setting or rising sun, producing colourful displays.

Even thinner and even rarer, *noctilucent clouds* sometimes form at amazingly high altitudes, high above the ozone layer at around 60 kilometres (35 miles) above sea level. They can become visible at dusk once the Sun has set over most of the atmosphere but still illuminates the upper levels. We are still on Earth but these clouds have a slightly alien feel to them.

Fig.5.8 The variety of clouds, on Earth and Mars. Image credit: NASA

Fig. 5.9 Noctilucent clouds from International Space Station. Image credit: NASA

The type of compound likely to condense into clouds depends on the temperature, and varies from planet to planet. Earth has water clouds[10]. Carbon dioxide clouds sometimes grace the Martian sky, and Venus is shrouded in sulphuric acid clouds. On the giant planets, successive cloud decks are made of different compounds as the temperature increases in the deeper layers.

The sequence of condensation temperatures for common molecules indicate which clouds can form on which planet. These molecules are, from colder temperature to hotter:

Methane	–161 °C
Ammonia	–78 °C
Carbon dioxide	–33 °C
Water	0 °C
Sulphuric acid	340 °C

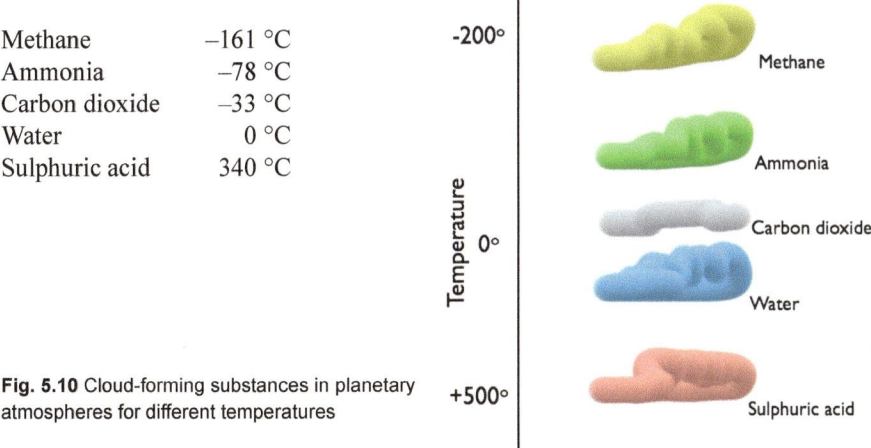

Fig. 5.10 Cloud-forming substances in planetary atmospheres for different temperatures

[10] Sometimes we also refer to "dust clouds" or "smoke clouds", which are not formed by condensation but by the presence of solid grains that make the air opaque. In this book we call this "haze" or "dust storms". They are also an important component of atmospheres, we have met them on Mars in Chapter 2 and will encounter them again on hot Jupiters in Chapter 6.

Why are these substances forming clouds and not others? Once more this is a consequence of the abundances of elements in the astronomer's periodic table. The atmosphere of giant planets is full of hydrogen, so the dominant form for the three abundant elements C, N and O is their hydrogen-rich versions H_2O (water), CH_4 (methane) and NH_3 (ammonia). It follows that these are the dominant sources of clouds on those planets[11]. Sulphur can also participate through NH_4SH, ammonium hydrosulphide, which forms clouds just below the level of ammonium clouds on Jupiter.

Jupiter is too close to the Sun for the temperature to drop below the condensation point of methane, but it does exhibit substantial ammonia and water clouds, whereas methane clouds readily form on Uranus and Neptune.

Circulation and weather

Clouds are a big part of what we call the "weather". On Earth as in other planets, clouds make the atmosphere visible, and allow us to follow its currents and circulation. The most spectacular clouds in the Solar System are undoubtedly on Jupiter, and there is apparently no limit to the variety of patterns that can play out on the King of planets. Indeed, Jupiter features so many clouds that they tend to blur into a psychedelic tapestry and make it difficult to visualise its atmosphere in three dimensions.

Cloud patterns are easier to interpret using infrared radiation which traces heat, and since clouds seen from space block out our view of the warmer regions below, the higher they are in the atmosphere, the darker they appear (weather satellites use this trick to monitor clouds on Earth during the night).

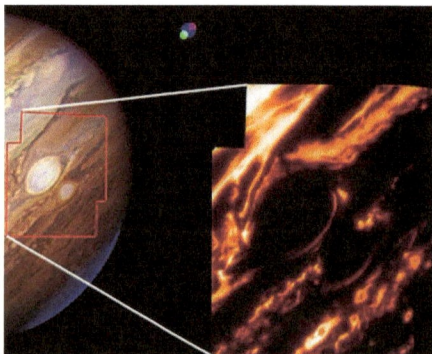 

Fig 5.11 Jupiter clouds in visible and infrared light. In infrared, the bright patches show deeper, hotter regions between the high clouds. Image credit: NASA

Fig 5.12 combines two views of Saturn, one in visible light, the other in the infrared. Visible light shows a smooth greenish diffusion from haze high up in the atmosphere, whereas in the infrared images, the patchy clouds below are visible, arranged in bands as on Jupiter, with the higher decks of clouds appearing as dark patches and the lower ones in bright red.

---

[11] Do clouds in gas giant planets smell bad? Ammonia and methane have something of a bad name. Ammonia is the smellier part of urine, and is the body's way to get rid of excess nitrogen. It smells foul, but is not very toxic. Methane is odourless but highly explosive. It is the main component of commercial natural gas. Cooking gas has a smelly compound added to it to help people detect leaks, so methane is often considered smelly by association.

Patterns of circulation

Why are the clouds on Jupiter and Saturn arranged in narrow bands? The difference between the global aspect of clouds on Jupiter and on Earth illustrates a profound relationship between rotation and atmospheric circulation.

Some science museums feature a large globe filled with coloured fluids, which the visitors can spin. If you rotate it very gently, nothing happens at first, then the fluids inside the transparent sphere start spinning with it. Push the sphere a bit faster, and some large motions begin to form, and in due course they majestically swirl across the sphere.

Push it faster still, and individual vortices appear, then the whole sphere starts looking like a satellite weather map of our planet.

Then, if you send it spinning even faster, something spectacular happens. At some point all poleward motion stops, and the currents arrange themselves in horizontal bands. The transition is sudden and startling. Push it a bit faster, and a new band appears out of nowhere.

Finally, if you give some irregular nudges to the globe, eddies form between the bands and evolve in fascinating patterns of waves, spirals and tentacles. The resemblance between this simple sphere and what is actually going on in Jupiter is striking.

This experiment illustrates some profound features about the circulation of planetary atmospheres, and why the main factor in determining the large-scale structure of the circulation is how fast the planet rotates on its own axis.

There are broadly three possible regimes for the motion of a fluid on a sphere as the rotation increases:

i. If the rotation is slow compared to the size of the planet, the fluid can move across the whole face of the globe, in large swirls and gentle waves. This can be seen on Mars, where dust storms propagate from the hemisphere closest to the sun to eventually cover the planet from pole to pole. Hot Jupiters are also in this regime.

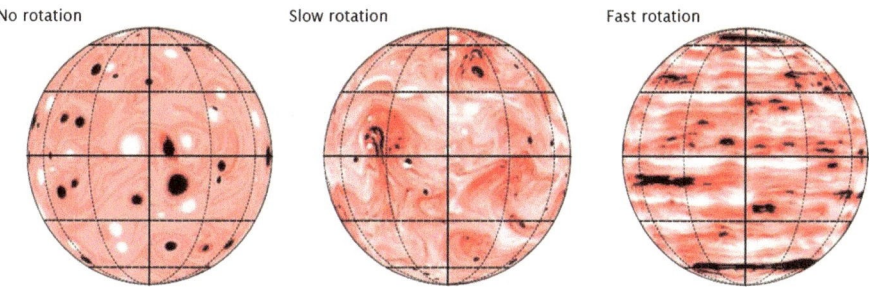

Fig. 5.13 Three regimes of atmospheric circulation. (Peter Read, after Yoden et al. 1999; Hayashi et al. 2000; Ishioka et al. 1999; Hayashi et al. 2007)

ii. As the rotation becomes rapid enough for the Coriolis effect (ice-skater/

rotating platform –see Chapter 1) to twist the currents before they can cross the whole planet, the circulation regime moves to a state dominated by a limited number of large vortices, arranged in a few bands lying parallel to the Equator.

Earth is an example of this regime. We do not see it so well because clouds cover only a fraction of the Earth and tend to avoid the tropics, but Earth's circulation follows six clear bands. The "bending distance" for weather patterns is approximately equal to the width of the bands, so that cyclonic patterns span a few thousand kilometres each. At any given time, about half a dozen high-pressure/low-pressure systems can be seen around each hemisphere.

Fig. 5.14 Earth's clouds. At any given time, the Earth's atmosphere shows about half a dozen weather systems in the mid-latitudes. Image credit: NASA

Fig. 5.15 Atmospheric circulation near the pole of Saturn. Image credit: NASA

iii) Finally, when the rotation is too fast for the fluid to travel polewards, the Coriolis effect becomes dominant and the currents fall into narrow horizontal bands. This is the state of Jupiter and Saturn.

Jupiter weather bulletin

The bands in the third type of circulation are far from static. Astronomers have been charting Jupiter since the time of Galileo (the man, not the probe), and it is clear that coming to understand the weather on this gas giant takes some patience, because the timescales are longer than on Earth.

The famous Great Red Spot is thought to be a slowly-evolving giant storm, which might last a few hundred years. In 2009, an even faster sign of evolution was observed: one of the bands, the large dark band just north of the Red Spot, simply vanished. Some whiter clouds seem to have formed at a higher altitude and quickly spread around the band.

Could space meteorologists have seen it coming? Can they explain it? The answer is the same as for weather on Earth, and brings us to one of the most challenging topics of modern physics.

Fig. 5.16 Jupiter's clouds. The different colours correspond mainly to different heights in the atmosphere. Notice the bluish regions, where gaps in the clouds allow us to see deeper into the planet. Image credit: NASA

Fig. 5.17 Jupiter at two different times. Note the disappearance of the main Southern band. Image credit: NASA

The role of turbulence

Eddies, twirls and swirls; the flow in planetary atmospheres is generally *turbulent*. Turbulent flow is a technical term used in physics to describe one of only two fundamental types of fluid flow, the other being laminar flow. The classic example of the difference between the two is that of cigarette smoke, which first rises straight up in a laminar manner, then abruptly breaks into turbulent swirls and curves. But most of us are now acquainted with another, more impressive transition from laminar to turbulent flow: when we are comfortably installed in an airplane, that seems to glide through air so smoothly that we could build a house of cards on the tray in front of us, and suddenly the whole plane jumps and rocks as if hit by sandbags from all sides at once. Outside there is not a cloud in sight, and nothing in the air looks different. The plane has just crossed from a laminar to a turbulent section of the sky.

Turbulent flow is one of the most intractable problems in physics. As Hungarian mathematician Theodore von Karman put it: "There are two great unexplained mysteries in our understanding of the universe. One is the nature of a unified generalised theory to explain both gravity and electromagnetism. The other is an understanding of the nature of turbulence. After I die, I expect God to clarify the general field theory to me. I have no such hope for turbulence".

What makes turbulence so intractable is related to the science of "chaos", phenomena that are fundamentally unpredictable, as often illustrated by the possibility that a butterfly flapping its wings in Brazil can, in principle, trigger a hurricane in Russia. Turbulent flow is chaotic, not in the sense of total disorder, but in the mathematical sense that small causes can produce immense effects.

Not all butterflies provoke storms, but the connection of large effects with potentially tiny causes makes it very difficult to study such processes with the usual tools of physics. The science of chaotic behaviour, sometimes also called complexity theory, is used to investigate the causes of avalanches and earthquakes, another field of study where it is possible to identify danger spots and likely events, but not to predict when and where anything will actually happen.

In a planetary atmosphere, turbulence means that large-scale properties, such as the red spot on Jupiter, or the speed of the global winds, will depend on the smallest scales, the weak interactions between particles in the tiniest of swirls, or a local knot in the magnetic field. It also means that there is really no way to predict in detail what the swirls and eddies will look like, or how they evolve.

Sometimes the best we can do is sit back and enjoy the view.

Fig. 5.18 Turbulence in a fluid-dynamics lab tank. Image credit: P. Burge

Chapter 6
Hot Jupiters

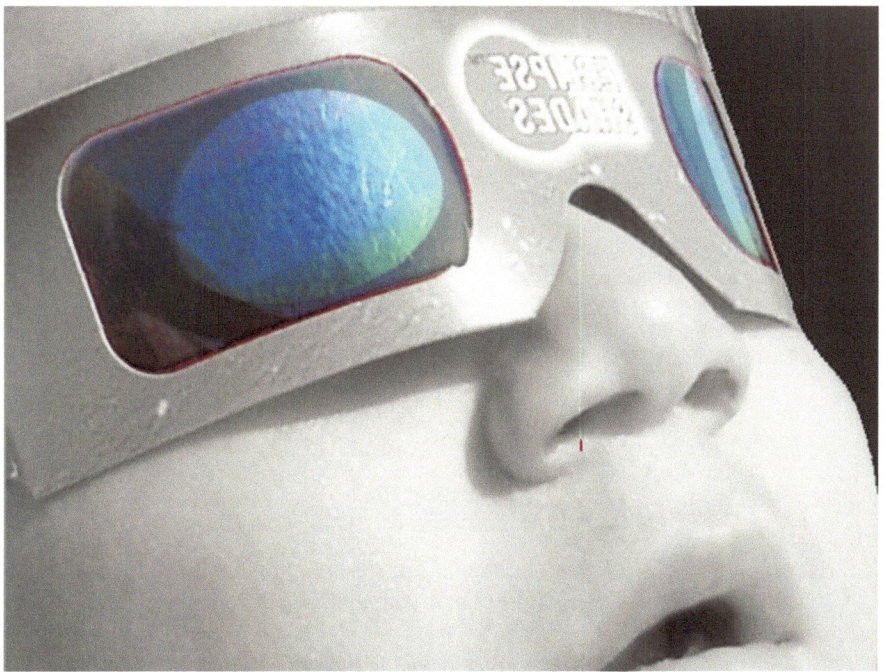

Gas giant exoplanets

Ever since the first planets around stars other than the Sun were discovered in the 1990s, we have learned quite a lot about exoplanets, with more than a thousand of all types and sizes now detected.

It turns out that planets are a very common by-product of the formation of stars from interstellar gas. When a new star is formed, more often than not it is accompanied by a host of planets, comets and asteroids. The best current estimates, by the NASA Kepler space planet-search mission, is that planets of the size of Earth are common, and number in the billions in our galaxy.

For the largest and most amenable to observation among the exoplanets, we have a chance to become personal, collecting not only their names and addresses, but also some of their intimate details and life history. One category of planets in particular has been the subject of intense scrutiny since the early 2000s, the *hot Jupiters*.

Hot Jupiters are similar in mass and composition to our Jupiter and Saturn. Like stars, they are mostly made up of hydrogen and helium, with a few percent of heavier materials such as water, rocks and metals. The difference is that they orbit much closer to their star than any planet in the Solar System, so close that they complete one orbit – their "year" – in only a few days, whereas Jupiter takes 11 years to circle the Sun.

Surveys suggest that about one in every 100 stars have a hot Jupiter close to them. This is surprisingly common, because before hot Jupiters were discovered, no planetary scientist predicted that gas giants would be found so close to their host star – the typical orbital distance of hot Jupiters is 1/20th of the Earth-Sun distance.

Hot planets

The proximity of the star is a crucial factor for a planetary atmosphere, as the differences between Venus, Earth and Titan illustrate. Venus is a scorching inferno, Earth a pleasant paradise (well, some of it), and Titan a frozen tangle of methane lakes, mostly because of their distance to the Sun.

Fig. 6.1 Size of the star 51 Pegasi seen from its hot Jupiter companion, and the Sun from Venus, Earth, Mars and Titan (from left to right).

Hot Jupiters receive enormous amounts of starlight, hundreds of times more than the Earth, which brings their atmospheric temperature above one thousand degrees Celsius.

Moreover, because their rotation is locked to their orbit, this heat is received only on one side of the planet, the "day side". The other side faces the dark cold of space.

The main problem for hot Jupiters' atmospheres is therefore dealing with this very simple issue: how to transport these massive amounts of heat around the planet, away from the infernal day-side towards the dark side.

Hot Jupiters cannot do that simply by spinning on themselves like the planets in the Solar System, their stars would not allow it.

When a gas planet orbits close to its much heavier star, it is welded into a teardrop shape by the gravitational pull of the star. If the planet spins faster on itself than it orbits around the star, the tip and bulge of the teardrop experience a strong pull that tends to bring them back towards synchronised rotation. In the same way, the Earth's

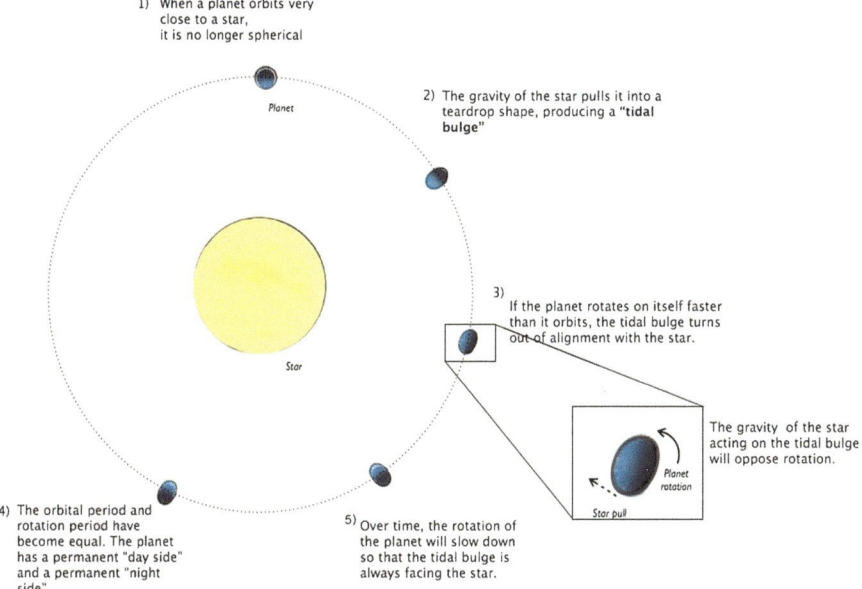

1) When a planet orbits very close to a star, it is no longer spherical

Planet

2) The gravity of the star pulls it into a teardrop shape, producing a **"tidal bulge"**

3) If the planet rotates on itself faster than it orbits, the tidal bulge turns out of alignment with the star.

Star

The gravity of the star acting on the tidal bulge will oppose rotation.

Planet rotation

Star pull

4) The orbital period and rotation period have become equal. The planet has a permanent "day side" and a permanent "night side"

5) Over time, the rotation of the planet will slow down so that the tidal bulge is always facing the star.

Fig. 6.2 Tidal synchronisation: when planets orbit close to their parent star, the pull of the tides on the planet synchronises the orbit with the rotation, so that the planet always presents the same side to the star.

Moon pulls the oceans out of shape and towards it twice every day on Earth, which is why this process is called *tidal* for exoplanets as well. Over time, the pull of the star wins over the spin of the planet, and the planet is forced into synching its spin with its orbit, and forever lining up its teardrop shape towards the star.

This is what happened to our Moon: it used to spin rapidly on itself, but over a few million years the gravitational pull of the Earth has slowed it down so that its heavier side now always faces towards us. Its orbital and rotation period have become identical, at 28 days. This is why we always see the same face of the Moon. An observer on the Moon, by contrast, sees the Earth spin on itself every day, because the gravitational pull of the Moon has not been enough to slow down the Earth sufficiently[12].

The temperature difference between the day side and the night side of hot Jupiters can be so strong that, in order to bring the heat from one side to the other, jet streams are pushed to velocities higher than the speed of sound (which is around 5,000 kilometres per hour in a hot Jupiter atmosphere, because sound travels faster in a lighter, hotter gas). When an atmospheric jet breaks the sound barrier it can create a shockwave. As the supersonic wind blasts into more slowly moving currents it cannot push them out of the way fast enough (the speed of sound is the speed at which "pushes" propagate through an atmosphere), causing the shockwave. In other words, the star's heat makes the atmosphere blast away from the day side!

What pattern do these jets follow? One could imagine a petal-shaped circulation, with the winds moving out in all directions from the centre of the day side (the point directly under the starlight), by the shortest path towards the night side. But although

---

[12] Although it did slow it down quite a lot. We now know that the Earth was rotating faster in the past. Geologists have found out that the duration of days has lengthened over the eras. For instance, a day was about 21 hours long in the Cambrian era. This is inferred from measuring how many days there were in a year from sea shells that grow a bit every day, and faster in the summer than in the winter.

the planet is always presenting the same face to the star, it is still spinning on itself – with a period equal to its orbital period. The rotation of the planet will, as on other types of planets, make polewards motion of currents more difficult than motions along latitude lines, so the atmospheric circulation will generally settle into global eastward jets, like on Venus.

These eastwards currents have been observed on one hot Jupiter, "Isis"[13], that we mentioned in the Preface.

Fig. 6.3 Petal-like circulation versus super-rotation – two possible ways to bring the heat from the day side to the night side of a hot Jupiter. Free-for-all winds rushing away from the hottest point, or global eastward currents.

Because of their high temperatures and the constant swirling of their atmospheres that whip up the planetary interiors by injecting heat into them, hot Jupiters can expand to sizes much larger than our own Jupiter, sometimes even to the point of slowly leaking gas into space. Osiris in the Pegasus constellation is surrounded like a comet by a gigantic tail of gas lost from its atmosphere. It is 1.3 times the size of Jupiter. The largest exoplanets we know reach twice the size of Jupiter, which is about eight times the volume, and larger than some diminutive stars.

A hot Jupiter tour

Let us embark on a spaceship and travel to the two nearby hot Jupiters that we know best, the ones that I have been calling Osiris and Isis, with official names HD 209458b and HD 189733b.

To travel to Osiris, we head for the constellation Pegasus in the northern sky, then straight ahead for 150 light-years or so, with occasional corrections to keep the star in our sights. On the scale of our galaxy, we are still in the solar neighbourhood. Not only have we not left the block, we are just in another room of the same building. As we draw nearer to the star, we notice a tiny speck of light glistening on its side. A mere ten star-diameters away from it, Osiris displays its splendid crescent, a dark, purplish glow basking in its host's glaring light.

Aspects of Osiris

The parent star of Osiris is a common sort of star, similar to the Sun except for a slightly higher temperature. Osiris is one of the "bloated" gas giant exoplanets, even larger than Jupiter. It orbits twice a week around its star, twenty times closer than the Earth is to its Sun. Its peculiar dark purple colour is due to the absorption of light by sodium – which swallows most of the incoming starlight, starting with orange.

The atmosphere of Osiris does not show the banded structure of Jupiter and the

---

[13] On the topic of planet naming, see notes at the end of the book.

other outer planets of the Solar System. Because of tidal locking its rotation period has to be identical to its orbital period, about three and a half days. In Chapter 5 we saw how a slow rotation implied wide-scale, planet-embracing weather patterns, and this is what we expect on Osiris.

The atmosphere of hot Jupiters is so hot that none of the molecules that we have encountered before in planetary clouds will be able to condense. Methane, ammonia, water, sulphuric acid, carbon dioxide, all these are far above their evaporation point, so they cannot condense into droplets or crystals to form clouds. All these molecules dissolve in a hot, transparent gas.

Could the whole atmosphere remain transparent, or could the whole planet be like a gigantic drop of water hovering in the sky, with the stars in the background showing through? This does not happen because forming clouds is not the only way for gases to absorb light.

The other way, that we know from measurements but for which we do not have a good intuitive grasp, is direct absorption of light by atoms. This is how ozone intercepts the harmful ultraviolet light from the Sun in the stratosphere, or how chlorophyll in plants captures sunlight to store energy.

In the hot gas atmosphere of Osiris, most of the incoming sunlight is captured by sodium and potassium.

Atoms capture photons by using their outer electrons as fishing nets. The more loosely their electrons are connected to the nucleus, the easier for them it will be to catch photons. In most atoms the electrons are too tightly attached to the nucleus to allow them to absorb photons of visible light. Only higher-energy photons, in the ultraviolet part of the spectrum, have sufficient energy to reach the inside of the atom. The exception is the *alkali* elements. These are the atoms in the first row of the Periodic Table of the elements, the first two being sodium and potassium. Their inner electron layers are full, and they have just one solitary electron in the outermost layer. This is the most favourable configuration for the absorption of light by an atom: the last

Fig. 6.4 Sodium gloom.

electron will be only loosely bound, available for other functions such as catching light or hooking to another atom.

In fact, the alkali elements are known chemically for their propensity to share their solitary electron very easily, which makes them chemically very active. They are eager to react with elements at the other extreme of the table, which lack only one or two electrons to complete their outer layer, like oxygen or chlorine. Such a reaction readily occurs between sodium and chlorine, yielding NaCl, table salt.

Another use of sodium that is related to the looseness of its outer electron is its use in city street lighting, the source of the slightly gloomy orange glow of modern suburbs. Sodium atoms can absorb visible light, and according to the laws of quantum physics that means they can also emit it. In a sodium lamp, sodium vapour is heated electrically until the outer electrons produce photons at the specific wavelength of sodium, orange.

In elemental form, sodium and potassium are light metals that look light grey, a bit like aluminium, except that they are soft as clay and easy to cut with a knife. Both elements are extremely reactive with water. Put into contact with H_2O, they steal the water's oxygen in a fast reaction. Pieces of sodium dropped in water burn and fizz until there is nothing left; potassium explodes instantly.

In the hot, transparent atmosphere of Osiris, sodium and potassium atoms capture most of the incoming light from the star. Paradoxically, this means that transparent-air hot Jupiters are very dark when viewed from space, because very little of the starlight will make its way back into space. Osiris is about as reflective as a piece of coal.

Also, somewhat paradoxically, the planet's colour will be "anti-orange". Since the atmosphere will catch orange light first, the planet will glow in the complementary colour, an ambiguous purple tinge.

Fig. 6.5 Sunset on Isis.   **Fig. 6.6** Sunset on Osiris.

Alien sunsets

We also know quite accurately what sunsets look like on Isis and Osiris. The colours of the setting star in the sky of the planet are exactly what is measured when collecting the spectrum during a transit, which was done for both planets with the Hubble Space Telescope, and the precision of the data is sufficient for a translation into colours as perceived by the human eye.

I have attempted this translation for these two planets. On Isis the sunset looks like a glorious sunset on Earth, on a very clear day with some dust in the air. This is because in both cases, Rayleigh scattering is the dominant mechanism. On Earth the scattering is caused by molecules and air-borne dust in the air. On Isis the Rayleigh scattering is thought to be caused by silicate dust. One key difference with a sunset on Earth is that the "Sun" appears much larger from Isis, because the planet is so much closer.

Osiris by contrast has a sunset that looks truly alien. The star is white outside the atmosphere since its temperature is close to that of the Sun. It then acquires a bluish tinge as it sinks deeper, because the absorption by sodium removes the red and orange from the starlight. Deeper down, Rayleigh scattering by the molecules in the atmosphere starts scattering the blue part of the spectrum as well, so that the only frequencies that are able to squeeze past are green, then murky brown. Outside the star's disc the atmosphere has a faint glow in its upper parts due to re-emission in the sodium lines; then it become bluer because of the Rayleigh scattering.

Rock and iron clouds

At 1,000 degrees Celsius and above, the temperature on hot Jupiters is much too high for carbon dioxide, water or methane to exist in liquid or solid form, but other substances can resist such temperatures. As a rule of thumb, heavier molecules evaporate at higher temperatures, so to find these refractory compounds we have to move down the Periodic Table into the realm of rocks and metals.

The most abundant of those are silicate rocks and iron, which will form tiny droplets when the temperature of the gas in a hot Jupiter cools, for instance when currents enter the obscure side of the planet. These droplets of glass (amorphous silicate) and metal can form clouds. On hot Jupiters, lava just rains out of the air, and glass storms lash out at sunset.

After Osiris, the next destination of our imaginary spaceship is Isis, a planet slightly cooler and shrouded in such glass clouds. We will approach this new giant planet and send a smaller capsule into its atmosphere, just like the gallant Galileo probe plunged into the clouds of Jupiter in 1995.

Travel to Isis

We now head towards the star HD 189733, 63 light-years from the Sun. Seen from Earth, the parent star of Isis is close to the spectacular Dumbbell Nebula, and we could take it as a landmark (skymark?) in our trip; but seen from Osiris the star appears in the direction of an entirely different region of the sky.

As we come closer, it becomes apparent that the host star of Isis is very different from the Sun, it emits a deep orange light. Eclipse shades reveal that its surface is sprinkled with dark star spots, which slowly drift across its face every few days.

Star spots

Stars have atmospheres too! Boiling, violently turbulent mixtures that bubble and burst from the inside to evacuate the intense heat of the nuclear reactions taking place in the core. One key difference with planetary atmospheres is that the intense heat – five or six thousand degrees Celsius – has stripped many atoms of one electron. As a result, the gas is a hot plasma that carries electric charges and responds to magnetic fields. On the intense boiling surface of the star, the magnetic field is tweaked and twisted in complex and ever-changing patterns, that channel the flow of plasma. In some regions the magnetic field becomes strong enough to prevent the hotter, lower layers from bubbling up to the surface. This produces temporary regions with lower temperatures – the star spots. Seen from the side, star spots are also regions where the stellar atmosphere is shallower.

Fig. 6.7 The atmosphere of the Sun. Image credit: NASA

Aspects of Isis

As on Osiris, powerful eastward jets transport the heat from the day side to the night side of the planet, whizzing past at 2,000 kilometres per hour. The temperature map of the atmosphere of Isis, measured with the Spitzer infrared space telescope, shows that the night side is heated to 700 degrees Celsius, proving that a large amount of heat is blown from the day side.

Because Isis spins slowly (its rotation is locked by tides), the weather features in the atmosphere are swirling across the whole planet. None of the tight banding and eddies of Jupiter here, more like the steady, obstinate flow around Venus.

Our spaceship edges closer to the planet and settles into orbit around it. Unlike the gas giant planets in our Solar System, Isis has no rings or moons to embellish it. The gravitational perturbation from the nearby star would rip them away in a short time.

Let us now complete one orbit around the planet. The day side is blindingly bright. Because the star is so close, the planet is a hundred times brighter than the full moon and eclipse shades are essential to look at it. Most of it is a deep featureless expanse, but some denser dust clouds occasionally trace the eastward currents. An oval hurricane can faintly be seen near the North Pole.

As we plunge towards the night side the star sets behind the planet, and everything is abruptly plunged into darkness. We can remove our shades and as we let our eyes adapt, one by one the outer stars appear in front of us. As we turn back to look at the planet a ghostly vision greets us. The night side is glowing with a deep, vibrating red hue, with swirls and tentacles tracing the paths of the hot currents flowing in from the day side. The raging wind is so hot that it produces the characteristic red glow of cinder and hot metal, a thermal emission that all bodies produce with a wavelength corresponding to their temperature. The Earth's thermal emission is completely in the infrared, and without special infrared goggles it is impossible to perceive even on the darkest of nights. Around 700 degrees Celsius most of the radiation is still in the infrared, but some spills into the visible red so the glow becomes perceptible to the naked eye.

Fig. 6.8 The bright side (left) and dark side (right) of Isis. The bright side is dark blue, possibly because a haze of glass droplets reflects blue light from the star and sodium gas absorbs red light. On the dark side, the hot currents blowing eastwards from the day side emit a red glow because of their high temperature.

The beauty and majesty of the night side of Isis at close range rivals that of Jupiter, in a different category. It has none of the elaborate bands, swirls, eddies and colours of our giant companion, but a single majestic broad-brush spectral theme.

And suddenly, as we approach the edge of the night side, the host star HD 189733, rises again and floods everything with intense light. The star is red in colour because it is markedly cooler than the Sun (merely 4,800 degrees Celsius compared to the Sun's 5,500 degrees Celsius). HD 189733 is also fainter than the Sun, but in our story we are so close that the brightness at this distance is many times stronger than the Sun seen from Earth. Another amazing view, never seen on Earth: a flood of red light many times brighter than sunlight at noon, a stunning view that may remain imprinted forever in our blinded retina if we don't put our eclipse shades back on very fast.

Taking the plunge

What we can learn about the atmosphere while remaining in orbit is limited. What is the haze made of? What lies underneath? Let us send a brave little probe covered with instruments to do this work for us. It seems safer to stay in the orbiter and record the results; the pressure and temperature conditions down there are not welcoming, and with the gravity of the planet it would be nearly impossible to crawl back out again.

Dropping from orbit, the probe falls towards the planet at such speed that it will cross the whole upper atmosphere in less than a minute, until friction slows it down. As it flashes by like a shooting star, the instruments quickly record the colour of the sky – light blue, no surprise there – and gathers spectroscopic snapshots that measure the composition of the atmosphere.

Compared to the previous hot Jupiter that we have visited, the colour of Isis is strikingly different. No more sodium anti-orange here, the planet is profoundly blue, as blue as the deep blue sea. This is not due to oceans, but to a cover of haze and dust that scatter the incoming starlight. An atmosphere that scatters light will acquire a deep blue colour, whatever the source of the scatter, provided the particles doing the scattering are very small – like atoms and molecules in the air and sea. That is why the clear sky is the same blue as a clean sea.

Deeper down, as the probe sheds the thermal shield which protected it during the initial slow-down and opens its Kevlar parachute, it enters the zone of the haze or cloud of glass. The instruments work frantically to gather as many recordings as possible in the shortest time and send them back to orbit. Is the haze made of enstatite, a magnesium-rich silicate that produces shiny, sparkly grains? Or is it olivine, a dark silicate, common in Earth basalt, that would produce blacker clouds? Or could it be iron droplets? The droplets slowly rain down (if that is the correct term) but as they reach even warmer layers inside the planet they evaporate again.

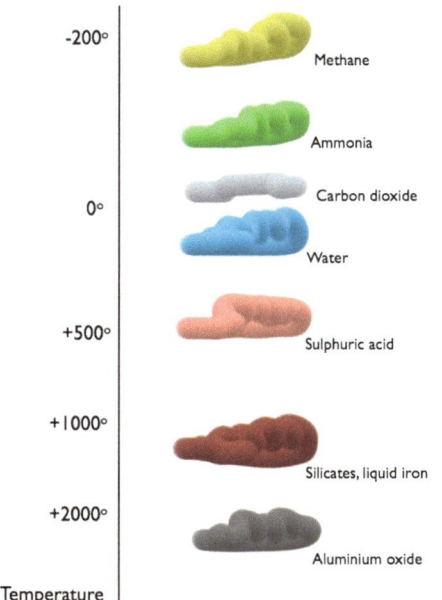

-200°

Methane

Ammonia

Carbon dioxide

0°

Water

+500°

Sulphuric acid

+1000°

Silicates, liquid iron

+2000°

Aluminium oxide

Temperature

Fig. 6.9 Cloud-forming substances in planetary atmospheres for different temperatures. This is an extension of Fig. 5.10 to higher temperatures. Above 1,000 Kelvin, droplets and grains of rocks, salts and metals can form clouds.

The probe has now reached a level where the pressure is equal to one atmosphere, like at sea level on Earth. The temperature is close to 1,000 degrees Celsius however, and the searing heat starts creeping through the probe's thermal insulation.

How hot?

Temperatures above forty degrees Celsius are unpleasant for humans. In the dry air of a sauna people can endure temperatures up to 100 degrees for short periods of time. Kitchen ovens usually reach 220 to 250 degrees, and the temperature around an ordinary flame is around 300 degrees.

Intense fires – for instance when a whole room is on fire and reaches what is called the flashover point at which everything burnable spontaneously ignites – reach around +1,000 degrees Celsius. Temperatures in this range are the maximum possible for combustion of carbon compounds (wood, houses, oil, natural gas) with oxygen.

Fire-fighters use *proximity suits* to resist an ambient heat of 260 degrees Celsius. The tougher category of *entry suit*, designed to step straight into the flames, provides protection up to 1,000 degrees Celsius for short periods.

Fig. 6.10 A "fire entry suit" designed to spend short moments in contact with fires up to 1,000 degrees Celsius. Image credit: Meikangco.com

The thermal tiles coating the bottom and front of the Space Shuttle resist temperatures of up to 1,600 degrees Celsius for the duration of the re-entry, just a few minutes.

Most metals start melting at temperatures between 1,200 and 1,500 degrees Celsius, with the exception of a few soft metals like mercury, lead and tin which have much lower melting points, and some hardy metals like tungsten which can withstand more heat.

Thus, relatively cool hot Jupiters like Isis have atmospheric temperatures which a fire-fighter in the best protection suit would resist for only a few moments, and in which copper or gold would soften rapidly. In the atmosphere of even hotter hot Jupiters, the suit would fry in an instant, and iron and steel would evaporate.

Time for a glass of water…

…and time for our probe to let go of its parachute, and drop faster before the instruments melt down. It is now becoming very dark around the probe, and the star, a very deep red, has almost vanished in the murky sky. That is where compounds like water, carbon dioxide and methane should be most clearly detectable in the infrared. Quick, robot, take a few spectra and send them back to the base! Also fire off a few Kevlar balloons, floaters equipped with transducers, to measure the wind speed, the eddies and the turbulence in this chaotic and stifling atmosphere.

The temperature and pressure have now exceeded the resistance of the probe, and its metallic carcass slowly sinks down into the dark, deeper levels of the atmosphere of the planet, until it reaches temperatures that will melt and vaporise any metal.

Fig. 6.11 Profile of the atmosphere of a hot Jupiter (based on present understanding of HD 189733b, "Isis") compared to Earth on the same pressure scale.

Interlude
Observations

Of course, dropping a descent probe into the atmosphere of a hot Jupiter is a planetary scientist's fantasy, and will certainly not happen any time in the foreseeable future. Nonetheless, the details described in our imaginary journey to Osiris and Isis are based on observations made in the past decade from our distant vantage point back here in the Solar System. The direction and speed of the winds, the trail of escaping gas, the light of sodium, all have been measured with telescopes. But when we can't even see the surface of Pluto from Earth with the Space Telescope, it seems difficult to imagine how we could observe the atmospheres of distant exoplanets. If it is such a challenge merely to detect them, how can we be studying their atmospheres from so far away?

Exoplanets are indeed vastly more distant than the outer planets in the Solar System. Saturn is a bit more than one light-hour away from us, Neptune is four light-hours away. The distance to the closest star is three light-years, and the nearest gas giant exoplanets that we know of are dozens of light-years from us. At that kind of distance they appear no larger than a coin lying on the surface of the Moon, and their glow is fainter than a distant galaxy.

This is not the only problem. Space telescopes and giant telescopes on the ground can now detect and resolve extremely faint galaxies at the other end of the Universe. The real problem that astronomers face when investigating exoplanets is the proximity of their stars, which are obviously so much larger and brighter than any planets they might have. Clever tricks are needed either to avoid the glare of the host star to get a glimpse of the planet, or, like in martial arts, to use the strength of the host star against itself in order to reveal the planet.

Among the methods used to detect extra-solar planets, one of the most powerful techniques is to look for *transits* – this is when a planet happens to cross the disc of its host star. Two such events were observable from Earth in 2004 and 2012 when Venus passed in front of the Sun.

About twice per century, Venus crosses the disc of the Sun as seen from Earth. The event has played a key role in the history of astronomy. It has permitted us to measure the scale of the Solar System, therefore the size of the Sun and its distance from Earth, as well as the size of the other planets. The trick is to observe the transit of Venus

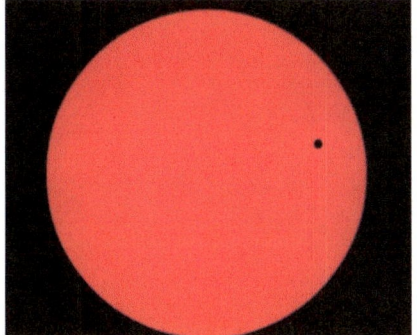

Fig. 6.12 and **6.13** The transit of Venus. Left: a photo of the 2008 transit. Image credit: Christine Wheeler. Right: commemorative stamp of Captain Cook's expedition to Tahiti for the 1769 transit.

from very different locations on Earth. The difference in the local time of the transit allows a measurement of the angle of the Earth as seen from the Sun, which provides the missing measurement to translate other observations like the apparent size of the Sun into real distances. Several expeditions were sent across the Earth to accomplish this key measurement for the two transits of 1761 and 1769, including the memorable expedition of Captain Cook to Tahiti.

In 2004 I was in Marseille on a bright summer day, using a small telescope to see our sister planet slowly crawl across the solar disc. Seeing an Earth-sized planet as such a puny dot on the Sun makes one realise the true scale of the Solar System. Early on a damp morning in June 2012, I found myself on a hilltop near Geneva trying to observe this again. It was a beautiful enough sight as the Sun slowly emerged over the lake, but low-lying clouds got in the way. This was rather a shame because the next transit of Venus is only due in 2117.

In the case of an exoplanet transit the stars are far too distant to be seen as discs. The trick to detecting transiting exoplanets is to monitor the total light of a large number of stars, and look for regular dips in the luminosity of some of them, which may be due to the transit of a planet.

For instance, in 2001 a telescope in the Chilean desert monitored a crowded patch of stars in the Milky Way to hunt for transits. One of the stars showed periodic dips that might have been due to a transiting exoplanet. But these measurements are difficult because the movements of air in the Earth's atmosphere blur the images and make it easy to confuse the light emitted by one star with that emitted by a neighbouring one.

With French colleagues we then used one of the European Southern Observatory's Very Large Telescopes in Chile – a behemoth with a primary mirror eight metres (26 feet) in diameter – to catch the transit with a higher precision. We also measured the motion of the star with enough precision to detect the tiny wobble induced by a planet orbiting around it. In that way we confirmed the presence of a transiting hot Jupiter

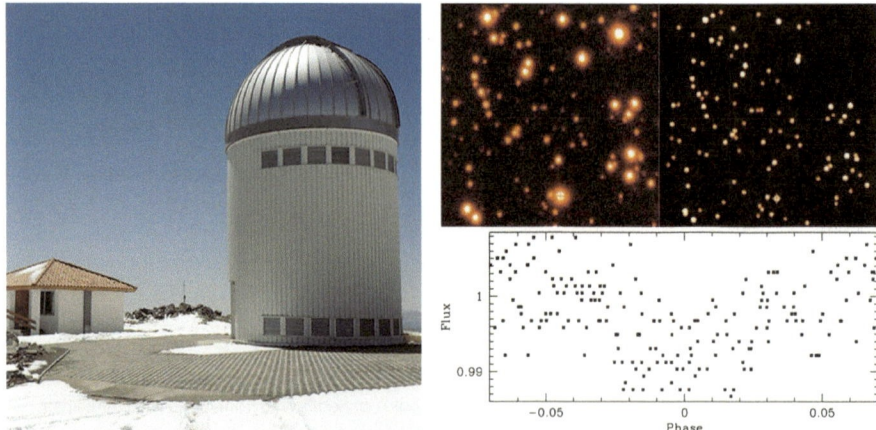

Fig. 6.14 The transit of OGLE-TR-132 with the Polish telescope (left) detected from the temporary dip in brightness (right). Image credit: Warsaw University Observatory at La Campanas

around the star, now called OGLE-TR-132b.

When a planet is detected by its transits, its size can be measured directly. The exact amount by which the stellar light dips when the planet crosses its disc is proportional to the surface of the planet. When combined with a mass measurement from the wobbles of the star due to the tug of the planet, we can get an idea of the density of the planet, and whether it is made primarily of gas or solid substances.

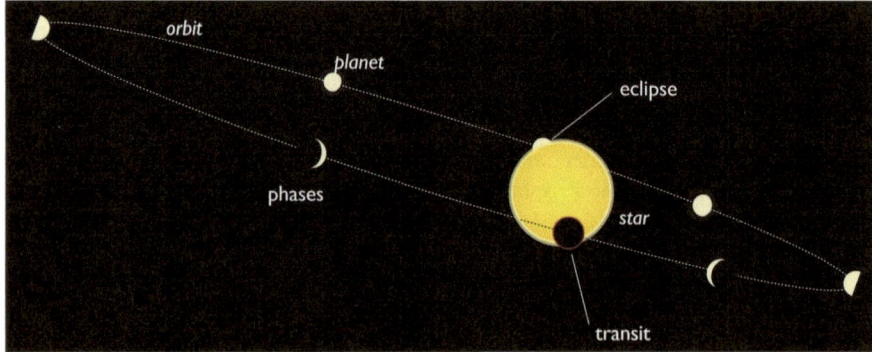

Fig. 6.15 The orbit and phases of a transiting planet.

It gets better: even if the transiting planet cannot be observed directly, with very sensitive measurements it is possible to measure its atmosphere. This is done in several ways. During the transit, if the planet has an atmosphere, some of the light from the star will filter through it before reaching the telescopes on Earth. This will make the planet look bigger in wavelengths where the atmosphere is more opaque, and smaller where it is more transparent. Which wavelengths of light an atmosphere absorbs or not reflects its composition and structure.

A second way to measure the atmosphere of transiting planets is to catch the

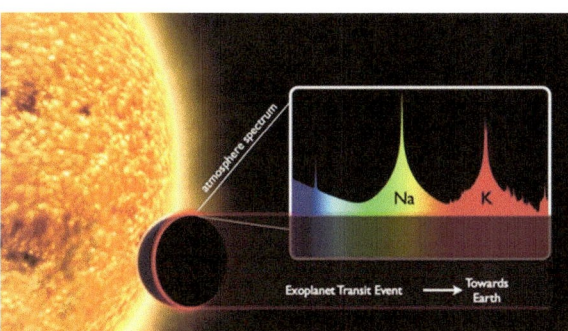

Fig. 6.16 Transmission spectroscopy: the spectrum of the atmosphere of a transiting planet can be reconstructed, because part of the starlight is filtered through the atmosphere during the transit. The spectrum of the planet in this sketch is dominated by sodium and potassium absorption, as expected for hot Jupiters without clouds. Image credit: David Sing

system at the other side of the orbit, when the planet disappears behind its star. The tiny resulting drop in total light will tell us how bright the surface of the planet was. Again, doing this at several different wavelengths yields a spectrum of the planetary atmosphere on the side facing the star. This difference in brightness is small: around 0.1 percent if it is a Jupiter-sized planet, to as little as 0.0001 percent for an Earth-like planet circling a Sun-like star.

A third piece of information is given by the phases of an exoplanet. This is the slight change in total light as the planet moves around its orbit, and presents different sides to us (like the phases of the Moon). Measuring the amplitude of the phases shows how hot the night side is compared to the day side, and, in the best of cases, enables the reconstruction of a temperature map of the whole atmosphere of the planet.

These three methods are at the source of most of what we know about the atmos-

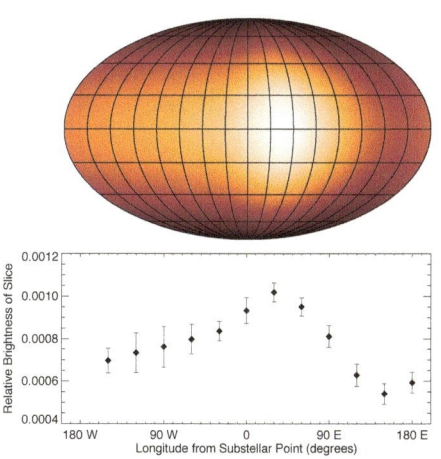

Fig. 6.17 Brightness map of the hot Jupiter Isis in the infrared, as inferred from measurements by the Spitzer Space Satellite. In the absence of winds or currents in the atmosphere the hottest point should be facing the star, on the Equator at zero longitude. Fast eastward winds shift the hottest points around 20 degrees in longitude. Image credit: Knutson *et al.*

pheres of exoplanets. They have been applied mainly using two space telescopes, *Hubble* for visible light, and *Spitzer* for infrared light. They could help us to understand the atmosphere of an Earth-like exoplanet.

On Earth, red light crosses the air more easily than blue light, which is why the setting Sun is red, and the sky blue. Infrared light is almost entirely blocked out by the water vapour present in the atmosphere, while ultraviolet light is intercepted by the ozone layer. To an alien astronomer, the Earth in transit would appear smallest in red, larger in blue, even larger in infrared (by about ten kilometres, the height of the troposphere), and largest of all in the UV (by 30–50 kilometres, the size of the ozone layer). From these values our alien observer could venture some informed guesses on the composition and structure of our atmosphere.

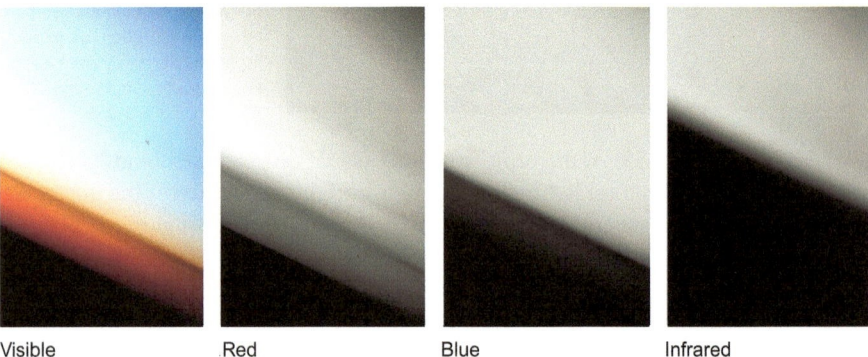

Visible   Red   Blue   Infrared

Fig. 6.18 Transmission spectroscopy of Earth's atmosphere from space. The Earth looks larger through a blue filter than a red one, because the atmosphere scatters blue light. It looks even larger in the infrared, because water vapour and CO_2 absorb infrared light. An alien astronomer could measure these features of our atmosphere during an Earth transit, even without seeing the planet, just by measuring the change in size in different wavelengths of light. Image credit: NASA/ISS

The transit methods work best for planets close to their host star, because the probability that a planet on a close orbit crosses the disc of its host star is much higher, and also because close-in planets are hotter, and therefore contribute a larger fraction to the light of the whole system.

As to the other end of the scale of orbital distances, for planets which are far enough from their host star (from five Astronomical Units like Jupiter to the edge of planetary systems at around 100 Astronomical Units), it is possible to conceal the intense stellar light and try to pick out the faint glow of the planet itself. Conceptually this is the simplest method, but it is technically demanding. Sophisticated tricks, like *interferometry*, are required to separate the firefly from the lighthouse, and this method is still in its infancy. Once the planet is identified in this way, however, it can be directly imaged and its spectrum measured without resorting to indirect methods.

These patchy and uncertain measurements are all we have today to ground our knowledge of exoplanet atmospheres. Just as it was a century ago for Solar System planets, some caution is due about our inferences and conclusions. The large brush strokes are probably correct but it is likely that many details will be revised as we gather better observations.

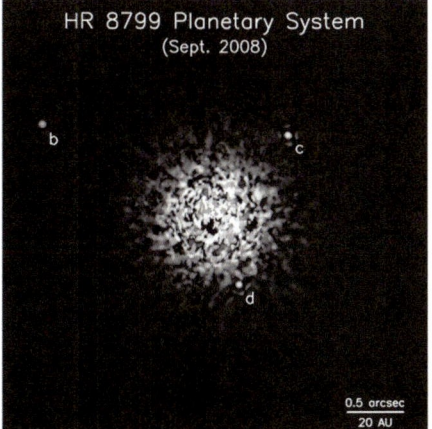

Fig. 6.19 An image of the HR8799 planetary system. The blinding light of the host star is suppressed, three planets are barely visible ("b" to "d"). Credit: C.Marois et al., NRC Canda

Fig. 6.20 Historical sketches of Jupiter and Saturn. Image credit: Agnes Giberne: Sun, Moon and Stars (1881)

Caution: observation-free zone ahead

The story is different for the next chapter. At this point, observations of the atmosphere of planets smaller than hot Jupiters are virtually nonexistent. What we can infer is entirely based on models, speculations and extrapolations. I have striven as much as possible to stick to inferences solidly grounded in known physics and astronomy, but if the past is any guide, some of the expectations presented in the next chapter will turn out to be completely incorrect, and the reality will be even more interesting and surprising than we thought.

Chapter 7
Terrestrial Exoplanets

As impressive as giant planets may be, they are not "our kind of planet". The ones we really want to know about are the smaller, Earth-like worlds. Not exactly like Earth, just close enough so we can relate to them, imagine ourselves there, roaming alien landscapes, the kind of places where we might even encounter life in some recognisable form.

About one hundred years ago, a common assumption was that the Universe and the Solar System itself were teeming with life. One of the reference books of popular astronomy at the time, by the French author Camille Flammarion, introduced its readers to likely Venusian jungles, ancient Martian cultures, and even ethereal inhabitants of Saturn's rings. Space was a natural stage for the extension of the great age of European exploration, and soon the galaxy was dotted with exotic alien tribes, egg-laying princesses and seven-armed beasts. All the medieval monsters and chimeras that had been chased from the terra incognita of Africa and Asia by explorers had retreated into outer space – where, indeed, they still are to this day.

But the planetary probes of the late twentieth century have blown a cold and dry wind on these alien dreams. Mars, then Venus, turned out to be totally hostile to life, and one by one planets and moons have been shown to be admittedly beautiful but totally mineral worlds. In spite of all the efforts of science communicators, planetary landscapes remain lifeless and emotionless, and become fascinating within the narrow confines of mineral elegance and abstract harmony.

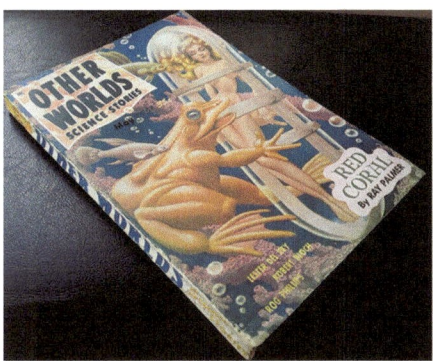

Fig. 7.1 Planetary landscapes do not measure up to fictional expectations. Image (right): NASA

But the alien dream is still there, filling books, movies and computer games by the hundreds, and its faint echo drifts up to the arid research institutes which pursue the study of terrestrial exoplanets. "Ocean planets", "super-Earths", "lava worlds", "sub-icecap oceans", these words all conjure up powerful emotions.

Let us now turn to these other worlds, as they are imagined by planetary scientists from the present, very patchy observations.

The planet gap

In the Solar System there is a large gap in mass between the smallest giant planet, Uranus, and the largest rocky planet, Earth. Uranus is 14 times heavier than our planet. Neptune has a mass close to that of Uranus, and Venus to Earth. But there is no planet in the mass interval between one and 14 times the mass of the Earth.

As we have seen on page 71, this division is caused by the snow line, the imaginary limit in planetary systems separating the inner regions that are too warm for water to condense into ice and snow, from the outer, colder regions. In the Solar System, the snow line is some three Astronomical Units from the Sun, between the orbits of Mars and Jupiter. The planets fall neatly into two groups, the small "dry" ones, made of rocks and metal, within the snow line, and the large "wet" planets, made mainly of ices and gas, beyond the snow line. There is about 15 times more water than rocks in the interstellar gas from which stars and planets are made. This is a reflection of the much higher abundance of hydrogen and oxygen than silicon and metals in the interstellar gas (see Astronomer's Periodic Table, page 70).

Because the snow line is present around all stars, we would predict a dearth of planets in the one to 15 Earth-mass range among exoplanets, as well.

However, the on-going census of exoplanets has shown that there are plenty of planets with masses between two and 14 times the mass of the Earth, and the expected gap is completely absent from the overall distribution of masses. Worse still, these planets do not particularly congregate at orbital distances related to the position of the snow line.

Fig. 7.2 Mass versus orbital distance diagram of known exoplanets.

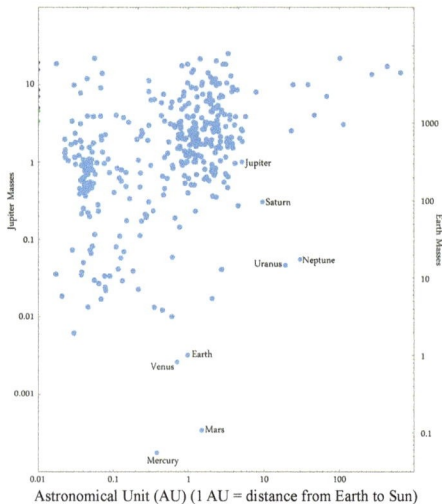

Astronomical Unit (AU) (1 AU = distance from Earth to Sun)

Armaggedons

Indeed, the most striking result from the hunt for exoplanets is that every plausible combination of planet size and distance seems to exist in some system somewhere. The exoplanets found so far show none of the order expected from the Solar System. They often orbit very close to their star – regardless of whether they contain ice or not – and generally follow more eccentric and tilted orbits than any planet in the Solar System.

This suggests that many planetary systems have messy histories, with plenty of near-collisions and orbital changes, which have shuffled the planets around and blurred the limit defined by the snow line. The Solar System seems to have been spared these gigantic rearrangements.

The science of exoplanets is still young and our views may shift as more planetary systems are discovered, but at present the understanding is that, after the dust in the disc had dissipated, most systems found themselves with more planets than they could accommodate. When planetary orbits are packed too close to each other, the planets repeatedly tug at each other gravitationally until their orbits become more and more eccentric. At some point this may bring two planets so close that the heavier one sends the lighter one on a wide loop, wreaking havoc on the whole system. Occasionally, planets can collide with each other and end up very close to the star – like hot Jupiters – or on the contrary very far away. They can even be swallowed by the star or be flung out of the system altogether.

After the birth of their system, planets find themselves locked in a gigantic game of musical chairs. There may be more than a dozen planets by the time the disc of gas around a new-born star dissipates, but room for less than ten. When a planetary system becomes unstable, two of its planets can crash against each other, forming a larger single planet, sometimes with moons formed out of the debris. However, the outcome can be more cataclysmic. If two gas giants interact so strongly that one of them spirals inwards towards the star to form a hot Jupiter, all the smaller planets in the system may well be lost in the process.

In the Solar System, each planet is about 1.5–2 times farther out than the previous one. This is about as tight as orbital dynamics would allow. If a new planet was inserted in the gap between two planets, it would make the whole system unstable. The one exception to the rule, the large distance between Mars and Jupiter, makes the point: it harbours cohorts of asteroids, the remains of a failed planet that was too close to Jupiter to form.

Upheaval in the Solar System

Our Solar System has escaped large-scale disaster. Its planets are on almost circular orbits, and they seem to have remained close to their birth places (in the sense that their present locations correspond to their internal composition, with rocky planets inside the snow line and ice-rich planets beyond).

However there is convincing evidence that we have had a very narrow escape. A team of scientists have reconstructed what may have been the most dramatic moments in the life of our planetary system. Here is the story.

Around 38 hundred million years ago, when the Solar System was seven hundred million years old, its giant planets were closer to each other than they are today, and Neptune and Uranus were in reverse order, with Neptune on the inner side, i.e. closer to Saturn. Over time, with the millions of small gravitational interactions between the planets, Saturn edged closer and closer to Jupiter, until it became too close for comfort. Jupiter began pulling Saturn into a more eccentric orbit, until its point of nearest approach became dangerously close to it. At some point the whole system snapped, Saturn was sharply pushed away by Jupiter, in turn sending the two smaller outer planets into wildly unstable orbits. Neptune was flung out to almost twice its former distance from the Sun while Uranus was bowled over so that it now spins almost upside-down. The myriad of minor planets crowded at the outer reaches of the Solar System (of which Pluto is the best-known example) were scattered all over the place, so that some were captured as satellites, some crashed into the giant planets, and some were sent inwards to fall onto the Earth or the Moon.

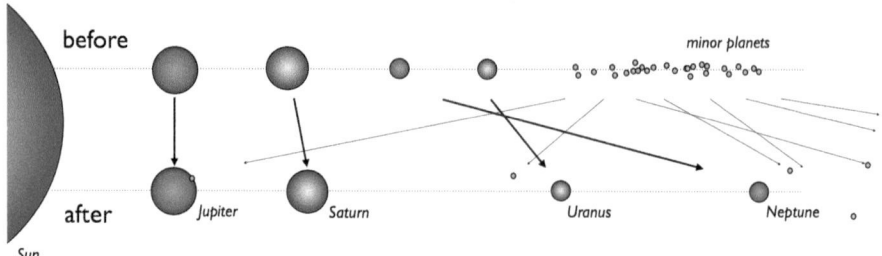

before

minor planets

after   *Jupiter*   *Saturn*   *Uranus*   *Neptune*

Sun

Fig. 7.3 Sketch of the Solar System before and after the upheaval of the "Late Heavy Bombardment" event, according to a plausible reconstruction. A resonance between Jupiter and Saturn spreads chaos throughout the outer regions of the Solar System, sending Uranus and Neptune to their present location, and sending a swarm of comets and asteroids everywhere.

This episode is known as the Late Heavy Bombardment. It is recorded by many huge craters on the Moon. It is "late" because it occurred about 700 million years after the formation of the Solar System, when such cataclysms were supposed to be a thing of the past. If this scenario is correct Jupiter almost killed his father a second time, destroying half of creation in the process.

Planetary dynamics offers the ultimate expression of the butterfly effect that we encountered earlier when talking about turbulent flow: how tiny changes in one place can propagate into the largest of consequences elsewhere. The dynamics of planetary orbits are very sensitive to this effect, indeed the butterfly effect and mathematical chaos were first described in relation to planetary orbits by the French mathematician Henri Poincaré in the nineteenth century. Poincaré was trying to prove that the configuration of planets in the Solar System is stable, that its orbits go on ticking forever like the model orreries built by astronomers. Instead, he got the first insight into the strange world of mathematical "chaos": he found that the tiniest of changes in the orbits could grow over time and evolve into unpredictable results.

Had the masses of Jupiter and Saturn been slightly different, the orbit of Saturn might have been perturbed beyond repair. The two giant planets could have collided, or Saturn could have been sent spiralling inwards to become one more "hot Jupiter", or outwards beyond Pluto. The first scenario would have spelt doom for the young Earth – already seven hundred million years old at that time and teeming with life, because all the inner planets would have been destroyed or scattered by a giant the size of Saturn spiralling towards the Sun.

Giant planets with orbits as circular as that of Jupiter are very rare among exoplanets. Less orderly outcomes of orbital interactions seem far more common. One could be tempted to speculate whether the uncharacteristic neatness of the Solar System is somehow related to our presence in it. It may be rare for Earth-like planets to be granted billions of years of quiet orbital cycles, so that organic life can evolve and prosper in peace. A single asteroid wiped out the dinosaurs, but that is next to nothing compared to rearrangements of whole planetary systems.

In case you are wondering, we seem to be safe this side of the Late Heavy Bombardment. Calculations about the future orbits of planets in the Solar System suggest that no further upheaval lies in store. There is a tiny chance that the orbit of Mercury will become more eccentric and start messing with the other rocky planets, but that is all. In computer simulations of the Solar System dynamics for five billion years in the future (the remaining lifetime of the Sun) 98 percent of cases have orbits remaining much as they are now. In 1–2 percent of cases, the orbit of Mercury becomes perturbed by Jupiter to such an extent that it either collides with Venus or falls into the Sun. In only one out of 2,500 of these simulations is the Earth strongly affected: the demise of Mercury bends the orbit of Mars until the inner point in its orbit crosses that of Earth. In a dramatic illustration of chaos theory, these different outcomes are obtained by changing the starting position of Mercury in the simulation by just one metre!

Composition of the atmosphere of terrestrial planets

Planets in the Solar system show that there are many possibilities for the atmosphere of terrestrial planets, indeed no two cases seem to be the same. Venus, Earth, Mars and Titan all have very different atmospheres, while Ganymede, Callisto and Europa – like our Moon – have no atmosphere at all.

As you may recall from Chapter 5, there are three possible origins for the gases which create atmospheres: they can be directly captured from the protoplanetary disc, expelled from the inside of the planet by volcanoes and lava flows, or added later on by comets and icy bodies crashing onto the planet.

In the first case, the planet must be heavy enough for its gravitational pull to retain the primordial gases even under the bombardment and collisions of the formation of the planet itself. Our Earth was not heavy enough for this, and must have lost all its initial gases early on, for instance during the giant impact that formed the Moon a few million years after the birth of the Solar System. However, planets around ten times heavier than Earth are able to retain their initial atmospheres, composed mainly of hydrogen and helium, like those of the Solar System's giant planets.

Alternatively, after the planet is formed, some lighter compounds trapped in the rocks may be released, in what is known as outgassing. A good example of this is planet Earth, where active volcanoes still spew out vast amounts of gases in the atmosphere and our atmosphere was mainly formed in this way. As convection in the Earth's mantle brings magma upwards, the lowering of the pressure causes chemical changes and rearrangements in the rocks that free volatile substances such as carbon dioxide, water and sulphur.

The importance of volcanism for the atmosphere of terrestrial planets may be surprising for most people, who live on large continental masses where magma and lava are remote presences underground, only providing heat for the occasional spa. But just as the rubbing together of continental plates is more familiar for people living in Japan, with weekly tremors as reminders, if you have visited places like the Yellowstone National Park in the United States, with fumaroles, bubbling lakes and geysers constantly blowing toxic fumes into the air over thousands of square kilometres, then the notion that the whole of Earth's atmosphere was ejected from the ground becomes more plausible.

Another source of volcanism is even less visible to us: the large chains of underwater volcanoes that form the ridges where tectonic plates surface from the interior of the Earth. The mid-ocean ridges are not only where new plates are formed, but also where the gases trapped in these plates are released into the oceans, later to make their way into the atmosphere.

On some small planets we can actually observe this process in a purer form. Io, the closest satellite of Jupiter, has hundreds of active volcanoes. Each time a volcano erupts it sends a plume hundreds of kilometres up into space, which forms a thin, temporary sulphur-rich atmosphere. But soon this tentative atmosphere is lost to space because the satellite's gravity is too low.

Volcanism and outgassing is not limited to lava and heavy gases as on Earth. On planets made mainly of ice, the main volcanic gas is water. This is "cryovolcanism": volcanoes of ice spewing out water vapour, a phenomenon that we have encountered on page 62 in the spectacular geysers of Enceladus.

Finally, comets visiting from outside the snow line can crash into a planet, vaporise on impact and enrich its atmosphere with volatile compounds such as water, ammonia, methane and carbon dioxide. This must also have happened on Earth to some degree, and some of the water in the oceans must come from comets.

These three kinds of atmospheres, originating in primordial gases, magma fumes or comets, look dramatically different, a difference that can be spotted at a distance of several light-years. The first type of atmosphere is very extended because the hydrogen and helium molecules are the lightest molecules. Instead of dropping by half every six kilometres as on Earth, the density of a hydrogen atmosphere drops by half every 40 kilometres.

Three categories of terrestrial planet atmospheres:		
Size	*Composition*	*Origin*
Extended	mainly hydrogen and helium	from interstellar gas
Medium	water, methane	from comets or water volcanoes (cryovolcanoes)
Compact	carbon dioxide, nitrogen, sulphur	from volcanoes

In one case, the planet Anubis[14] situated in the Ophiuchus constellation, transit observations suggest that the atmosphere is compact. The size and mass of this planet suggest that it is mainly made of water, so a water vapour atmosphere would be a natural explanation. At this stage the evidence is mostly indirect, though a compact atmosphere is inferred from the fact that no spectral lines were detected during the transit in spite of many attempts with telescopes on the ground or in space. In a hydrogen-helium atmosphere these lines would be large enough to be seen.

There may be a fourth category of atmospheres. Some planets are so close to their star that all the light elements will have been blown into space by the high temperatures in the atmosphere (a higher temperature means that the molecules move about faster, and faster motion means an easier escape from the planet's gravity). If the planet has a rock mantle, its lava surface would exude silicate and metallic vapours that may form a thin atmosphere of "rock vapour".

---

[14] Another name I've just made up, the planet is officially called GJ1214b.

Exploring the zoo of imaginary terrestrial exoplanets

What do terrestrial exoplanets look like, feel like, smell like? Planetary scientists are not waiting for more observations before they start exploring the zoo of possibilities. But before taking a stroll across the bestiary that they have created, we must pause for a little warning. The history of science in general, and exoplanet science in particular, has not been too kind to unbridled speculations or unconstrained models. Confident predictions later overturned by observations are too many to mention. Indeed, conversely, model predictions later confirmed by observations are so rare as to be counted on the hands of a Saturnian (which, according to Camille Flammarion, must have robust hands to deal with the higher gravity, implying a lower number of fingers than us).

There is good news, however; what the observations reveal is very often wilder, more varied, more colourful and more interesting than what the models predicted. Keeping this in mind, let's take a tour of the exoplanet zoo and stop briefly in front of a few exotic specimens.

Ocean planet

Fig. 7.4 Ocean planets: the dream.

First, we come across one of the water worlds. This one has the correct temperature for water to be liquid, so we will call it Aquaworld, to distinguish it from its cooler (ice) and hotter (supercritical) water-rich cousins. To imagine what such a planet would be like, we start from a climate model of the Earth, and raise the level of the oceans until

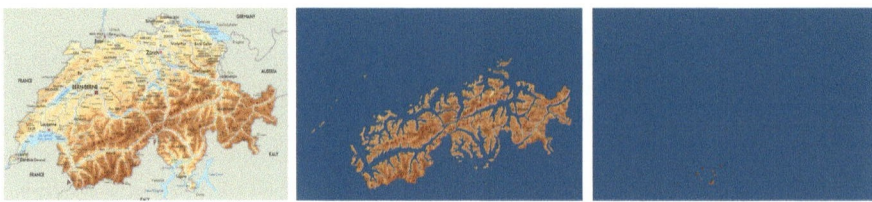

Fig. 7.5 Switzerland now, with sea level 1,500 metres (5,000 feet) higher, and four kilometres (13,000 feet) higher.

all the continents are covered.

On Earth, the continents divert the flow of currents in the oceans, forcing them along a complicated loop which snakes around the whole world, transporting heat from the Equator to the poles in the process.

Without continents, oceanic currents would be much simpler. They would behave like the atmospheric currents, with mostly eastwards currents because of the planetary rotation (on a planet the "east" is by definition the direction of rotation).

That would make the poleward transport of heat less efficient, leading to more deeply frozen poles than on the present Earth.

In fact, this phenomenon is visible on Earth: near the North Pole, the climate is relatively benign, with people living as high north as latitude 70 degrees, because the shapes of the continents channel the warm Atlantic currents all the way into the Arctic Ocean. The relatively pleasant town of Tromsø in Norway is at +69 degrees 40 minutes, the somewhat tougher town of Barrow in Alaska is at +71degrees 17 minutes. In the south, by contrast, the sea currents can circle the whole Antarctic continent using the Drake Passage between South America and the Antarctic Peninsula. This is why the south polar region remains frozen from around –60 degrees southwards. The southernmost towns on Earth are Puerto Williams and Ushuaia in Tierra del Fuego at a latitude of –54 degrees south. Incidentally, the opening of the Drake passage by tectonic motions about 41 million years ago was one of the drivers for the ice ages that have seized the Earth at various intervals during the last few million years. An apparently local event such as the opening of Drake's passage has caused global climate change!

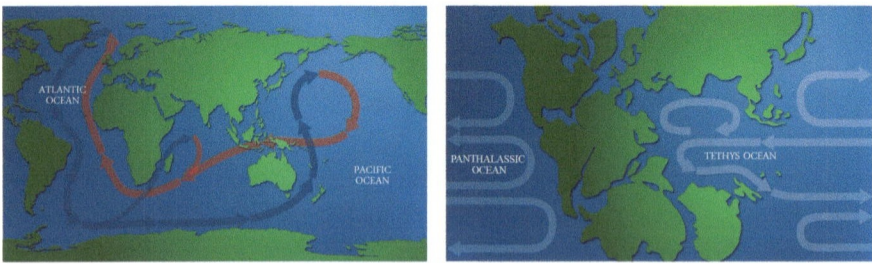

Fig. 7.6 Global ocean currents now, and in the time of Pangea, 200 million years ago.

Figure 7.7 shows, step by step, how simplified models of the position of continents can make us understand the circulation on hypothetical Aquaworld and on Earth.

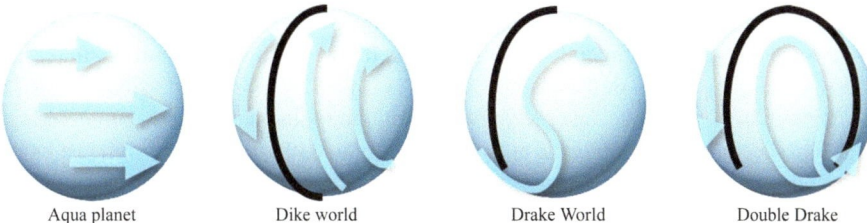

Aqua planet   Dike world   Drake World   Double Drake

Fig. 7.7 Ocean currents with four highly simplified distributions of continents. Present Earth broadly corresponds to the fourth case.

In Chapter 3 we mentioned the daunting engineering project of rendering Mars inhabitable. The comparison of the currents in our "Drake World" and a hypothetical "Dike World" suggest a way to render Antarctica inhabitable, which, as such grandiose projects go, may be easier than terraforming Mars. We "only" need to build a dike across the Drake passage, and the ocean currents will be diverted. The loop around Antarctica that keeps it locked in a cold polar pocket will be broken. The whole continent will heat up, most of the ice sheet will melt, and 14 million square kilometres of prime real estate will be made available to an eager human population.

The only work required is to connect the Tierra del Fuego at the tip of South America with the Antarctica Peninsula, a stretch of ocean 1,000 kilometres (600 miles) in length, with a dam of earth and gravel. This is some engineering challenge to be sure, but possibly easier than pumping a whole atmosphere into Mars.

Snowball planet

In its long history, planet Earth has explored configurations that are so different from each other that they could actually represent other planets. Several times in the past, all Earth's continents were gathered up in a single landmass. Soon after its birth Earth was probably covered by a single ocean without any emerging land. In a sense Earth has even travelled several distances from its host star, since the Sun has become markedly brighter along the eons, corresponding to moving the Earth 20 percent closer to its host star since its birth.

Fig. 7.8 Snowball planet.

Perhaps the most intriguing and unfamiliar state from Earth's past is the one known as *snowball Earth*. At several stages in its life, before the apparition of complex life-forms and the colonisation of land, Earth was seized by extreme versions of the Ice Age, with most of the ocean freezing over, and the global temperature plummeting.

This seems to have occurred at least twice, about 650 million years ago just before the appearance of complex animals in the "Cambrian explosion", and earlier around 2.2 billion years ago. These episodes lasted a few dozen million years before the climate started thawing again.

It turns out that the climate on Earth can find stability in two configurations: the present state, with relatively warm seas and continents which, thanks to the dark colour of water, rocks and vegetation, absorb enough sunlight to maintain the mean global temperature around +15 degrees Celsius; and a snow-and-ice covered state, where most of the sunlight is reflected back into space because snow, being white, absorbs very little light.

Both states are stable, which is a cautionary tale when trying to predict the climate of an exoplanet from models and calculations; even in a single case like the Earth, there are two perfectly valid solutions.

It is not clear what triggered the snowball states on Earth – nor whether they were global. Some part of the oceans near the Equator might have remained ice-free. We are not sure, either, what it was that jolted the planet out of these states. The best candidate so far is the greenhouse effect induced by the carbon dioxide released in volcanic eruptions. Normally, the oceans and rocks on the continents absorb as much carbon dioxide as the volcanoes produce, keeping the amount in the atmosphere constant. This is a delicate balance that can be offset by changes in the position of the continents, the brightness of the Sun or the chemistry of life. If too much carbon dioxide is captured, the greenhouse effect drops, and the temperature crosses the point at which snowfalls start sending the planet into a positive feedback loop: more snowfall means more sunlight sent back into space, cooler temperatures, and more snowfalls.

Fortunately for us, there is a longer-term negative feedback that provides an escape from snowball Earth: once it covers oceans and continents, ice prevents the absorption of carbon dioxide by water and rocks. Volcanoes can pierce through the ice though, and they keep pouring CO_2 into the atmosphere, until the greenhouse effect builds up sufficiently to melt the ice again. The feedback loop is now inverted, with more open seas meaning darker colours, more absorbed sunlight and a higher temperature melting more icecaps.

Terrestrial water planets positioned slightly further away from their host star than the Earth, or orbiting a fainter star, may become permanently blocked in this snowball state, creating beautiful but barren polar worlds.

If the planet is large and young enough, hidden oceans may subsist deep below the ice. In Antarctica, Lake Vostok (which is an enormous lake buried under several miles of ice) is kept liquid by volcanoes buried underneath the ice sheet, in spite of temperatures between –30 and –60 degrees Celsius at the surface.

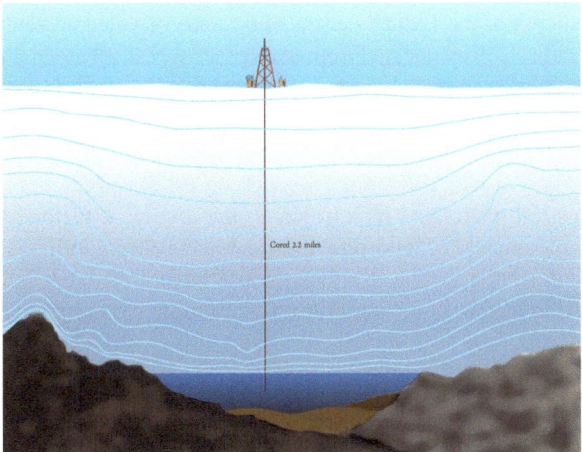

Fig. 7.9 Lake Vostok.

A snowball exo-Earth could keep a large buried ocean in liquid form under a frozen surface.

Fig. 7.10 Europa's surface, rafts of ice sheet on a buried ocean. Image credit: NASA

Steppenwolf planet

A planet could even keep an under-ice ocean warm in the absence of light from a star. A planet drifting in empty space, light-years from the nearest star, could still harbour dark but warm seas under the protective cover of a thick ice sheet, heated by underwater lava flows. There is enough residual heat in a planet the size of Earth, and heat generated by the disintegration of radio-active atoms, to keep its volcanoes going for billions of years.

Indeed, astronomers have found that planets drifting in empty space, far from any parent star, are not uncommon. They will have been ejected from their system by a heavier planet or after a close encounter with another star. The idea that such planets might harbour life with their dark interior oceans powered by underwater volcanoes is not that far-fetched. In the depth of the Earth's oceans, thousands of volcanic vents maintain living communities entirely disconnected from the surface world, all along the mid-ocean ridges. The catch is that even if such a planet is teeming with life, it is hard to imagine what kind of detectable signal it would leave on the outside of the planet.

Different Earths

The distance between a terrestrial planet and its host star is the dominant factor for a terrestrial planet, but other parameters of its orbital motion have a strong influence on its atmosphere and climate. An Earth-like exoplanet could have a different rate of spin, changing the length of a day, or a different tilt of its rotation axis, or a more eccentric orbit around its star.

Rotation speed

In Chapter 3 we saw how the speed of rotation affects a planet's climate. At one end of the spectrum slowly rotating planets have a single circulation pattern covering the whole planet. At the other extreme, fast rotating planets form narrow east-west bands. For rotation rates somewhere in-between, the patterns will consist of a few large zones. How would this affect a terrestrial planet? The main effect of rotation would be to prevent the efficient transport of heat from the equatorial regions to the poles. For a planet rotating up to twice faster than Earth, the difference would be slight. The transport of heat from Equator to Pole would be made more difficult by the increase in the Coriolis effect.

But at a faster rotation, with days becoming shorter than about ten hours, the atmosphere would switch to a banded structure like that of Jupiter, and heat transport towards the poles would become inefficient. The poles would remain extremely cold. During the long, dark winters, temperatures would probably reach the point of condensation of carbon dioxide (–78 degree Celsius), so that dry ice would start falling out of the sky in small crystals, a carbon dioxide snow.

Some planets would have become locked to their star so that their rotation period and orbital period would have converged to the same value, in other words the duration of days and years become the same (through the tidal mechanism that we encountered on page 85). Like hot Jupiters such planets have a permanent day side and a permanent night side, and the Sun never moves in the sky. If the planet has little or no atmosphere, the consequences on its climate are drastic, with extreme temperature differences between the two sides of the planet. But with an atmosphere, winds will carry the heat across the planet, and negate some of the day-night contrast. A thick atmosphere like that of Venus will negate the day-night difference almost entirely.

With a "medium weight" atmosphere, for instance an Earth equivalent, the contrast between the day and night side will not be entirely erased. This configuration has been studied by climate specialists, and they predict a mean temperature of +50 degrees Celsius near the Equator on the side of the planet always facing the Sun, and a mean temperature of –10 at the centre of the night side. Powerful westerly winds would bring some of the heat to the night side. The transition from day to night would not be a question of time like on Earth, but a distance covered in space. The Sun would never rise or set in any given place, but a long journey towards the east or west will make the Sun rise or fall in the sky.

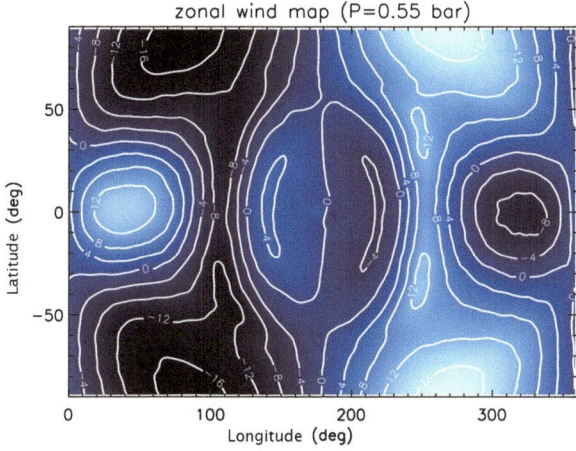

zonal wind map (P=0.55 bar)

Fig. 7.11 Plot of the wind speed from a climate simulation for a tidally-locked terrestrial planet. Image credit: Heng et al.

Climate on such a rotation-synchronised earth would be peculiar. The difference that we are used to associating with the north-south axis would be reported to east and west. The day side, around longitude zero degrees, would have a permanently hot climate, the centre of the night side, longitude 180 degrees, would be permanently frozen. But the regions near the edge (90 degrees east and 90 degrees west) could have very variable weather, depending on whether the main plume of hot air from the day side crosses the region or swerves past it.

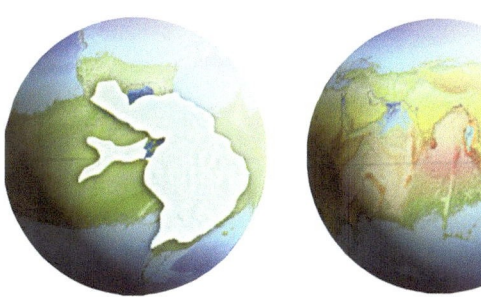

Fig. 7.12 Climate on tidally synchronised Earth. The densest part of Earth, situated in Asia, always faces the Sun. An ice sheet covers most of the Americas. Since the centrifugal force has diminished, the oceans have retreated towards the poles.

Axial tilt

The rotation axis of the Earth is tilted by 23 degrees relative to the Sun. The regions most exposed to sunlight oscillate during each year between latitude 23 degrees north and 23 degrees south, the Tropics of Cancer and Capricorn, and the North and South Pole spend half the year in darkness and half in constant sunlight. In temperate countries, the height of the Sun in the sky changes by 46 degrees between winter and summer. If the rotational axis of the Earth was aligned with the way it rotates around the Sun, each region would receive the same amount of sunlight throughout the year, and there would be no seasons. The poles would still be colder than the tropics because sunlight hits them at a low angle.

There is nothing special about this value of 23 degrees, some planets are tilted completely sideways (Uranus), so that their poles get more sunlight than the Equator, while other are flipped upside down (Venus), so that they orbit the Sun clockwise but rotate counter-clockwise. A planet's tilt is not even constant with time, but can vary under the influence of other objects. Earth's tilt varies by about one degree every few thousand years. It is relatively stable because the gravity of the Moon tends to prevent large swings in its position. Mars, by contrast, is thought to have oscillated between values as different as 11 to 49 degrees.

An Earth-like planet with an axial tilt higher than Earth's present 23 degrees would have more extreme seasons. We get a taste of this in arid, high-latitude regions like the Gobi desert, because dry air amplifies temperature differences. The temperature in Mongolia often rises to +40 degrees Celsius in the summer and plummets to –30 degrees Celsius in the winter.

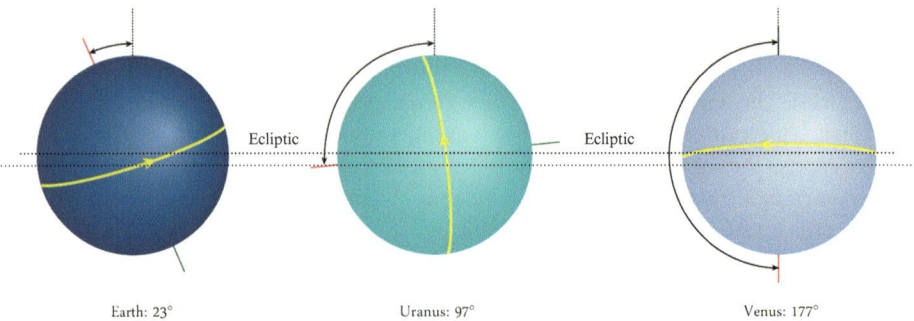

Earth: 23°   Uranus: 97°   Venus: 177°

Fig. 7.13 The axial tilt, or obliquity, is the angle between the rotation axis of a planet and the plane of its rotation around the Sun.

Beyond 53 degrees of inclination, the poles receive more heat during a year than the Equator does, reversing the usual hierarchy. On an Earth-like planet tilted at such a high angle, the warmest climate would be found near the poles in summer! The other hemisphere would get extremely cold in the winter.

The appearance of the Sun in the sky viewed from such a planet takes some imagining, and you'll see it better if you just try it out with an orange and an apple than by reading about it. Near the poles, the Sun would stay high in the sky for the whole summer, in a sort of permanent tropical noon, then slowly set in autumn and disappear for the whole winter, causing scarcely imaginable seasonal variations.

Closer to the Equator, the length of the day would more constant, but the seasonal variations extreme, with the Sun barely skimming the horizon in spring and autumn and climbing towards the zenith at noon in summer. Huge storms would form over the summer pole as the winds try to carry the heat away towards the darker hemisphere.

The climate and air circulation on such a planet would be profoundly different. One possibility explored by planetary scientists is that of a "pool planet", covered with ice and snow except in a seasonal opening of the sea at the summer pole. One can only ponder how life would adapt to such a climate. Life could still survive in the ocean under the ice, as in the Arctic on Earth, but all photosynthetic organisms would need to follow the pool of open water with the seasons.

Eccentric orbit

Johannes Kepler proved mathematically in the seventeenth century that planetary orbits follow ellipses, but the elongation of the ellipse can take any value. In the Solar System, the planets follow nearly circular orbits, but elongated orbits are common among exoplanets. In fact, the vast majority of exoplanets follow orbits that are more elongated than those of the Solar System. The nearby hot Jupiter Nephthys[15], some 200 light-years away in Ursa Major, boasts a staggering 93 percent eccentricity. This means that the planet is seven percent of the size of its orbit from the star at the closest point, and 93 percent at the farthest. In other words, its distance from the star varies by a factor of 14, and so does the diameter of the star as seen from the planet! Since apparent size varies with the square of the distance, it implies that its sun appears nearly 200 times brighter at the closest point than at the furthest on that planet.

Mars has the highest orbital eccentricity among Solar System planets – nine percent; this is sufficient to cause a large difference between the climate of its Southern and Northern hemispheres. The South polar cap, which spends its "winters" near the furthest point from the Sun, is much more extended than the North. The Earth's orbital eccentricity is 1.7 percent, which does not appreciably modify the seasons – remembering that the seasons on Earth are not caused by the variation of the distance from the Sun, but by obliquity.

An Earth-like extrasolar planet on a very eccentric orbit has another type of season. These "eccentricity seasons" are not due to change in the height of the Sun in the sky, but to the planet moving toward and away from its sun. On Earth they are imperceptible; on Mars they contribute to triggering global dust storms in the southern hemisphere, when the obliquity and eccentricity seasons coincide.

If the Earth's orbit, for instance, had an eccentricity of 50 percent, it would spend part of the year closer to the Sun than Venus, and another part further away than Mars! Such a planet might spend only a fraction of its time within the "habitability zone", the zone around the Sun where liquid water is stable on a terrestrial planet. Close to the star, surface temperatures would reach water-boiling levels, with lakes and shallow seas evaporated, clouds cleared by the searing heat and the air thickened by a photochemical haze. Then, after crossing the habitable zone, autumn would feature torrential rains pouring all the evaporated water back onto the surface; finally ice fields would rapidly cover all the newly re-formed seas for the winter.

While these conditions seem uncomfortable, would they actually make life impossible? Not necessarily. Even our own planet presents striking climatic variations that are difficult to adapt to. For one thing, it switches from sunlight to total darkness and back every 24 hours. Then the temperature can vary by more than 60 degrees throughout the year in a given location, and some regions see no sunlight for months on end. Life has evolved strategies to deal with these changes, most often by shifting to "intermittent" modes that allow it to wait out the difficulties. Animals sleep at night

---

[15] Real name HD 80606b.

or hibernate during the cold season, desert plants hide underground during dry spells, bacteria can freeze and thaw again. Maybe adapting to the "eccentric lifestyle" is not that much harder than coping with seasons and diurnal cycles. Moreover, oceans take a long time to heat or cool, so marine creatures remain sheltered from the wild swings of the surface climate.

Another feature of orbital dynamics makes the climate on a very eccentric planet even more interesting. Let us consider, for instance, a terrestrial planet on an orbit like that of Nephthys, with a year of 365 days like Earth but an eccentricity of 93 percent. The pull of the star near the closest point of its orbit is so strong that it will force the

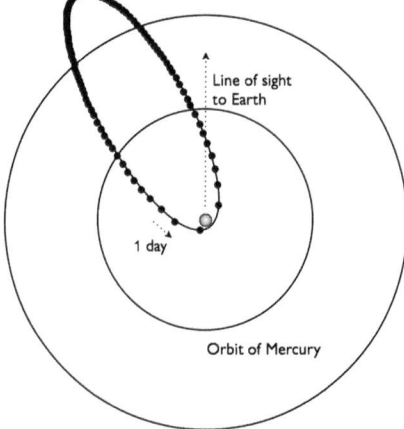

Fig. 7.14 The orbit of Nephthys, compared to the scale of our Solar System. The dots are placed one day apart on the highly eccentric orbit. Image credit: Gregory Laughlin

planetary rotation into line with the orbital velocity, so that at this time, the planet will always present the same face to the star. During the rest of the year, the planet will keep this rotation speed, but since it proceeds more slowly on its orbit as it moves away from the star, it will rotate faster than it orbits.

The consequences for the planet are spectacular. The "winter" consists of slow days, lasting about ten Earth days each, with the sun very slowly inching through the sky, then setting for five days, before appearing again. Every night, lakes freeze over, and during the mid-afternoon the sun finally becomes strong enough to melt some pools on the ice sheets until they freeze again ten Earth days later during the following "night". During the course of the year, as spring progresses and the midsummer solstice draws nearer, the climate becomes slowly warmer.

As midsummer approaches, the sun starts slowing down in the sky, the days lengthen so much that summer consists of only one or two month-long days. Come the summer solstice, the view is stunning: the Sun grinds to a halt in the sky, some 14 times larger and 200 times brighter than in the depths of winter, and stays there for hours, before slowly starting to move westward again. All life forms on the day-side of an Earth-like planet with such an orbit would probably dig themselves deep into the ground or sink into the depths of the sea to avoid being roasted by ultraviolet radiation. Gigantic floods would stream down the slopes of the continents, bringing down all the glaciers and icecaps at once. In tropical regions, as the ground temperature hits 100 degrees the rivers and streams would start to boil.

Then a few days later the air cools, and another perfectly normal year on that planet begins again. The dazed inhabitants come out of their summer dug-outs to celebrate the end of the sun festival.

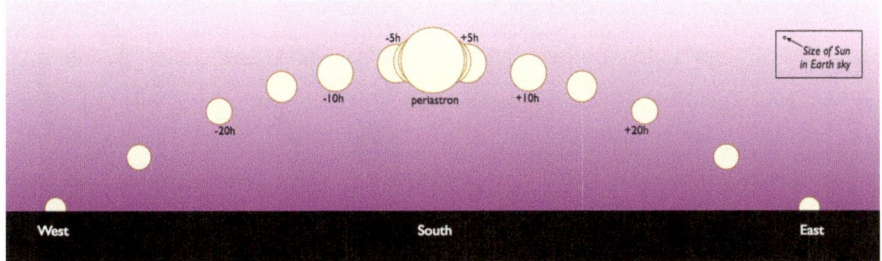

Fig. 7.15 Trajectory of the sun in the sky of an Earth-like planet with a very eccentric orbit, near the summer solstice.

Diamond planets

There is another way in which an Earth-like planet can differ from Earth. We have seen that the abundance of elements in the cosmos constrains the composition of planets. There are no 'copper planets' or 'salt planets'. Planets are basically restricted to three mixtures: rock and metal, ices, and hydrogen/helium.

There is, though, one intriguing possibility, the carbon-rich planets sometimes dubbed "diamond planets".

In interstellar gas there is more oxygen than carbon. Since oxygen is so reactive, it locks all the carbon into CO and CO_2. Then the oxygen left over is available to combine with metals to form rocks and ore.

But the composition of interstellar gas varies from place to place in our galaxy. Dying stars and supernovae produce different mixes of elements depending on how large they are, and since carbon is a close fourth behind oxygen in abundance, it is possible to imagine that some planet-forming discs could have more carbon than oxygen.

In an oxygen-rich planet like ours, metallic oxides dominate the interior. The vast majority are silicon oxides – silicates – what we commonly call rocks. But in a carbon-rich planet, all the oxygen would be locked up with the carbon, and there will still be carbon left over to bind with the metals and form carbon-only compounds. The planet crust would be rich in silicon carbide, graphite, coal and yes, deeper down at higher pressure, diamond. Silicon carbide is a dark, shiny rock, used in modern car brake discs for instance. It occurs very rarely on Earth, because there is generally enough oxygen around to oxidise it into silicates or carbonates. Diamond is the hardest mineral known to humankind and has surprising properties, not least of which is to separate people from vast amounts of money.

What would the atmosphere of a carbon-rich "diamond planet" be like? The major components would be carbon monoxide (CO) and methane (CH_4), the simplest carbon compounds with hydrogen and oxygen. The surface would consist of very dark rocks,

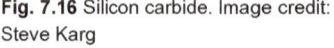

Fig. 7.16 Silicon carbide. Image credit: Steve Karg

Fig. 7.17 A good friend of diamonds. Image credit: Milton H. Greene

graphite, coal and silicon carbide, and a host of other compounds because carbon is chemically versatile, with occasional outcrops of diamonds emerging from the interior. Most striking would be the near complete absence of water.

Plate tectonics would probably be impossible on a diamond planet because of the toughness of carbon compounds, and the absence of the main plate lubricant, liquid water. The internal heat of the planet would have to be evacuated in hot spots, lava flows and large volcanoes, like on Mars. The melting point of silicon carbide is scarily high at 2,720 degrees Celsius, so the magma would be either much hotter than in Earth's volcanoes, or formed from some softer and rarer carbon components. Diamond and graphite are even harder to melt, with fusion temperatures above 3,500 degrees Celsius.

Some planetary scientists have been trying to explore the broad features of such worlds, but we are really out on a limb here. Planetary atmospheres are complex enough for model-makers and require constant supervision in the form of observations and empirical verification. Nevertheless, the image of a silicon carbide mountain range with diamond cliffs is hard to ignore.

Exo plate tectonics

In the Solar System, the intensity and importance of volcanic activity rises sharply with the size of the planet. This is a manifestation of the surface-to-volume, or "elephant and mouse" effect. The mass of a planet rises with the *cube* of its size, and so does the amount of heat it contains. The surface of a planet increases as the *square* of its size, as does the surface available for evacuating the heat. Smaller bodies therefore cool more rapidly, because they have more surface per element of volume. Elephants need huge ears to cool down, while mice have to move fast and eat a lot to keep warm. That is why the Moon can cool without any volcanism at all, losing heat to space by conduction of heat through its crust of rocks, like a radiator. Mars features a few large but very inactive volcanoes and lava flows. The surface of Venus, on the other hand, shows signs of intense volcanic activity, and on Earth chains of volcanoes line the boundaries of the moving plates.

Some terrestrial exoplanets are much larger than Earth. A natural expectation is that the amount of volcanic activity would keep growing with increasing mass. This would imply that "super-size Earths" generally have thicker atmospheres than Earth or even Venus. Another consequence is that, even in the absence of water, the crust could become warm enough and the convection of magma sufficiently intense to force surface plates to slide against each other. Since heat is evacuated by the lava at the plates' boundary, the best way to evacuate more heat is to create more boundaries. In that case, we would expect the plates to be smaller and more numerous than on Earth.

On such a planet the cycle of gases between the atmosphere and the rocks would occur much faster than on the terrestrial planets in the Solar System. It takes geological eras for the content of the Earth's atmosphere to cycle through oceans, continents and volcanoes, but with a faster cycle and a thicker atmosphere, there might not be enough time for the atmosphere to evolve on its own, so it would remain close to the composition of volcanic fumes.

At present, there is no measurement of the conditions in the atmosphere of heavy terrestrial planets, and the models have to be pushed so far out of their comfort zone to explore this question that they may be only marginally more reliable than informed guesses.

Now that we have wandered close to the realm of idle speculation, we might as well dip straight into it and talk about life on exoplanets, a topic on which the absence of empirical information is even more complete.

The search for life

Seen from the cosmos, terrestrial planets are tiny by-products of star formations, like crumbs left on the table. The reason we care so much about these planets is that we happen to live on one. Carbon planets, super-Earths, eccentric, tilted or synchronised planets; what we want to know is whether life could appear and thrive on them.

Life has had a profound effect on the atmosphere of our planet, so great, in fact, that the change could easily be spotted from space by alien astronomers. The presence of abundant oxygen in our atmosphere produces features in the spectrum of light emitted by our planet that could be picked up by a telescope dozens of light-years away. Indeed the composition of the atmospheres of Mars and Venus were measured in this way from Earth, one century ago. The ozone layer, formed by the action of sunlight on the abundant oxygen, absorbs ultra-violet light at specific frequencies, leaving its recognisable fingerprint on the spectrum. Alien observers could conclude that something unusual has happened to this particular planet, and, depending on their knowledge and prejudice on the topic, infer the presence of life from the suspicious abundance of substances as corrosive as oxygen and ozone.

To confirm this suspicion, alien astronomers may collect more measurements and spot the weaker signs of methane in the Earth's atmosphere. Methane and oxygen react strongly with each other (which is why cooking gas burns so well), so their simultaneous presence in an atmosphere implies a constant source producing each of them. Life on Earth does this, with some bacteria exuding methane and photosynthetic plants producing vast amounts of oxygen.

For a few decades now, finding both oxygen and methane simultaneously in the atmosphere of a planet has been "plan A" for scientists in the search for primitive life on Earth-like exoplanets. (In contrast, intelligent life in the Universe is being searched for by the SETI programme, listening in to galactic radio communication.)

Other stars are so distant that measuring the spectrum of a terrestrial exoplanet with sufficient accuracy would be an enormous technical challenge. In the 1980s, space agencies dreamt up a concept involving a flotilla of space telescopes linked to each other to function like a single giant telescope. This is what would be needed to collect sufficiently strong signals to be able to separate the planet from its host star at a distance of several light-years. In 1999, NASA administrator Dan Golding gave a moving speech showing a mock poster of an exo-Earth, with continent and ocean shapes outlined as if imaged painstakingly from an astronomical distance. "Can you imagine….? " he would say, "in around 30 years, such pictures could be hanging in classrooms." A powerful dream.

But twenty years on, "around 30 years" is still what it might take to achieve such a project, and the technical and financial hurdles have come more sharply into focus. The space interferometer Terrestrial Planet Finder project has remained on the to-do list for a generation; like those Post-Its that have been stuck on the fridge door for so long that nobody dares to take them away, but when they finally drop off, nobody puts them back.

Or maybe the next generation will. The project has not been shelved entirely, and it is still "plan A". An important step was taken in 2011, with the Kepler space mission measuring the abundance of terrestrial planets around normal stars. This abundance is high enough for some of them to be within reach of "plan A", with an Earth twin around every dozen stars or so. The second step is to find a few such twins around nearby stars, and projects are under consideration to launch a space mission to do this by 2020. Then, for the flotilla of probes, we may have to wait "around 30 years".

It is not only an issue of money and politics. There are also some interesting scientific arguments against focussing too narrowly on oxygen and ozone. Research on the history of the Earth has shown that our own planet has harboured oxygen-producing photosynthetic life for at least 3.4 billion of its 4.6-billion-year life. But for most of that period, no large amount of oxygen has accumulated in the atmosphere. Only in the past 500 million years or so has the oxygen concentration passed the one-percent mark before shooting up to the present 20 percent. And only at these high concentrations has an ozone layer appeared. Bacteria and algae produced oxygen for three billion years in a world that was able to absorb it immediately, without any accumulation in the atmosphere that could be detected from space. Given this, "plan A" seems risky. Even if there are several Earth-like planets in our neighbourhood inhabited by carbon-based life, 90 percent of these planets may still not display the giveaway signs of methane, oxygen and ozone.

"Plan A" has not gone away, in spite of the difficulties, partly because there is no obvious alternative. Our galaxy could be teeming with life forms that do not affect the atmosphere of their planet in a detectable way. But our kind of methane and oxygen-producing life is the only one we can think of detecting with realistic technology.

Living planets
Could that simply be due to a failure on the part of our imagination? Some scientists think so.

Let us return to what is special about the coexistence of methane and oxygen. The key point here is that such coexistence is a sign of *out of equilibrium* chemistry. Maybe, instead of looking for this specific sign of disequilibrium, we could look for disequilibrium in general.

The concept of equilibrium is central to chemistry. In a laboratory, molecules react with each other until they reach an equilibrium, specified by the laws of physics and chemistry. Create an out-of-equilibrium state by mixing two compounds, and they will react with each other to reach an equilibrium.

Outside interference can keep a chemical mix out of equilibrium, for instance the sunlight illuminating an atmosphere, or volcanic fumes being added to it. But these kinds of disequilibrium are nothing compared to what life is able to achieve. Methane in the Earth's atmosphere is about 10^{30} times more abundant than it would be at equilibrium; that is a lot of zeroes.

Maybe we should go back to the reason why oxygen is produced in the first place. Stealing energy from sunlight requires a sophisticated piece of biochemical engineering called photosynthesis. This involves swapping electrons between one molecule and another, using one photon of sunlight. The process requires an "electron donor", and the most common element, hydrogen, also happens to be one of the easiest to part with its electron. So photosynthesis uses hydrogen as an electron donor.

But hydrogen doesn't float around freely in the sea or in the atmosphere, so that photosynthesis has had to find a way of extracting it from another molecule – water. The water molecule, H_2O, is cleaved so that the two hydrogen atoms can be used as electron donors. The remaining free oxygen is released as a highly reactive free radical.

However, water is a notoriously stable molecule, so breaking it requires a lot of energy. Other hydrogen compounds are much easier to break, for instance hydrogen sulphide, H_2S. Some scientists think that photosynthesis may actually have started using these easier sources of hydrogen – H_2S is common in volcanic products for instance – then moved on to more difficult molecules over time. It could have taken hundreds of millions of years to evolve the trick of breaking the most abundant molecule in the ocean, and dealing with its toxic by-product.

One smart hypothesis is that using H_2S may not have evolved directly as a trick to capture the energy of sunlight, but as a way for bacteria living near underwater volcanoes to locate themselves relative to the source of heat. Living near an oceanic "hydrothermal vent" is tricky, because the temperature changes quickly from over one hundred degrees near the vent to four degrees Celsius in areas unaffected by its heat. An organism that could determine its position relative to the vent by detecting H_2S would be at a great advantage for survival. Since we now think that life may have started near these volcanoes in the abyss, far away from sunlight, this hypothesis provides a neat way of explaining the origin of photosynthesis by gradual steps, rather than one big jump.

The implication is that photosynthesis on other planets need not produce oxygen. It may be based on other processes such as sulphur chemistry.

Therefore what we could be looking for is not necessarily a methane-oxygen pair, but any sign of a global chemical imbalance in the atmosphere of a planet.

As of 2014, the next planned landmarks in the search for habitable planets and life include the James Webb Space Telescope, the successor to the Hubble Space Telescope, scheduled for the end of the decade, which could measure the atmosphere of a few Earth-like planets, as well as a smaller space telescope focussed specifically on measuring the signature of molecules in planetary atmospheres in the infrared. The largest telescopes on the ground have already begun the search for planets that can be seen directly – rather than only through the transit method. But reaching planets as small as Earth with this method may require the next generation of telescopes, behemoths with mirrors 30 to 40 metres (100–130 feet) across, instead of the current six–ten metres, that are planned for the 2020s.

Apart from detecting substances in the atmosphere, there are other things that can be learnt about an Earth-like planet, even by an observer too remote to make out any surface features. By monitoring the brightness and colour of an exo-Earth over several days, it is possible to reconstruct a broad-brush map of its surface, and measure how it varies because of changing clouds. Some researchers have used the *Epoxi* space probe, en route to a comet, to measure our own planet for a few days. Putting themselves in the position of alien astronomers, they have inferred the broad position of the oceans and continents from the evolution of the brightness and colour of the planet as its different parts rotated in and out of view during its day.

In terms of surface features and cloud coverage, Earth is strikingly different from other planets in the Solar System. Its surface is divided between oceans (which look dark blue from space) and continents (bright and brown), and about half its surface is covered by white clouds at any given time, in ever-changing patterns. By contrast, the global aspect of Venus is very uniform. Mars has dark and bright patches, occasionally covered by dust storms, but the daily changes in brightness and colours are much less sharp than on Earth.

A similar map of an exo-Earth would not tell us whether it harbours life or not, but it would go a long way to telling us what kind of world it is. Some specialists think that the existence of active plate tectonics on Earth – which is reflected in the separation between oceans and continents – is essential to the development of life. And the presence of clouds thick enough to indicate a healthy water cycle but not so thick as to shroud the whole planet constantly, would also be an encouraging indication.

We might be better off choosing a gradual, flexible approach to the detection of life on other planets, with each step depending on what we have learnt from the previous one, rather than the all-or-nothing gamble of "plan A". The Kepler mission has taught us that Earth-like planets are common. Now we can find some of them around nearby stars, and study them in an open-minded way, trying to characterise the diversity of

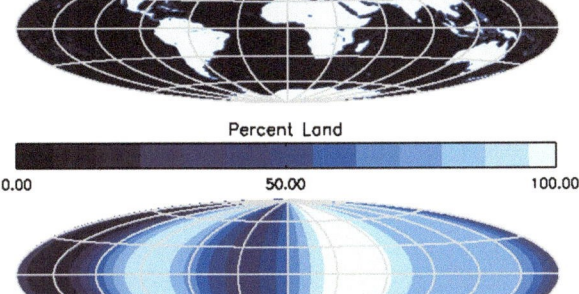

Fig. 7.18 The broad distribution of land and water on Earth reconstructed from distant observations by the Epoxi space probe. Image credit: Cowan et al.

their atmospheres. Then, if and when we have more definite suspicions about one or some of them, we could plan more targeted observations. Of course, each step in this quest is long, complex and expensive, but it is a quest that is as old as civilisation, so we might have to be patient.

Chapter 8
Back to Earth

There is a set of terrestrial planets that we can study much closer to home. They are separated from us not by space but by time. Our planet has been around for 4.6 billion years, about a third of the total age of the Universe (13.7 billion years). Throughout its long life, it has changed so much that, were we to be transported to one of its past incarnations, it would appear as alien as an exoplanet.

The Earth has been in turn a lava planet, an ocean planet and a snowball planet. The main constituent of the air has been successively hydrogen, water vapour, carbon dioxide, and finally nitrogen. The pressure at the surface has risen to hundreds of bars before falling again to the one bar to which we are accustomed.

Today we could say that the Earth is having some sort of mid-life crisis, because one species on its surface has been busy burning vast quantities of buried carbon and fossil fuels, thus abruptly increasing the carbon dioxide content of the atmosphere.

And the story is not over. We are only mid-way through the life of the Sun, and more dramatic changes to the Earth's atmosphere are in store.

Let us now journey through some of the stages of the Earth's rich history, which has been painstakingly reconstructed by a motley crew of geologists, geophysicists, biologists and astrophysicists.

But first, to tell the history of the Earth we need a more convenient unit of time than years, which are just too small. Planets evolve over hundreds of millions of years, the duration of geological epochs. There is no convenient name in English for hectomegayear, and we have to resort to a geological term: one hundred million years is an *Era*. Our planet is now 46 Eras old, in ripe middle age. The dinosaurs roamed the planet one Era ago. Earth's life expectancy is about 90 Eras, at which point the Sun will inflate into a red giant star and die[16].

Let us leaf through the family photo album and look at Earth at ages one, five, 24, 40 and 46 Eras.

Birth pangs

When the young Sun became bright enough to blow away the disc of gas that surrounded it at birth, it was left with a procession of planets. There were at least four gas giants far away, so hot that they shone like small stars. Five red-hot balls of lava orbited closer to the Sun. Starting from the Sun these unruly children were Mercury, Venus, Earth, Theia, then Mars.

Earth's atmosphere was made up mostly of hydrogen and helium captured from the disc of gas. But because these two elements are so light, they leaked out of the atmosphere into space at a steady rate.

Then one morning, around 4.4 billion years ago, young Earth was hit by the Mars-sized planet Theia. The cores of the two planets merged, but enough magma was thrown out by the collision to form the Moon. This event was so violent that no rock would have been left unturned, or more accurately unmelted, so that the Moon-forming impact can be taken as the true birth date of "our" Earth. Amazingly, the whole catastrophe lasted less than a day, the most memorable day in Earth's history.

The Moon impact suggests why the atmospheres of Earth-like planets around other stars may be very different from that of Earth, even at the earliest stages. The details of the last impacts can completely modify the outer crust of a rocky planet and the amount and types of gases available to form the atmosphere.

Early in its life, the Earth was a "lava planet", entirely covered with an ocean of liquid molten rock.

Some exoplanets remain covered with lava all their life, because they orbit so close to their star that the temperature remains hot enough to melt rock. But the nascent Earth cooled quickly. Soon, bits of crust formed, like the crust on lava flows in volcanic places like Hawaii. Because solid rock can hold less gas than molten lava, the gases contained in the rocks escaped. These volcanic gases, mainly water, carbon dioxide, and sulphur, formed the second atmosphere of planet Earth – after the atmosphere of hydrogen and helium left over from the disc of gas around the Sun. For a few million

---

[16] Accidents may happen to shorten this lifetime, such as a very close encounter with another star that perturbs the orbits in the Solar System, but such encounters are extremely rare because the space between the stars is so vast. There is also the small possibility of the orbit of Mercury becoming unstable, see page 107.

Fig. 8.1 26 April, 4370426528 BC, 3:12 AM: moon-forming sequence of events. Image credit: James Symonds

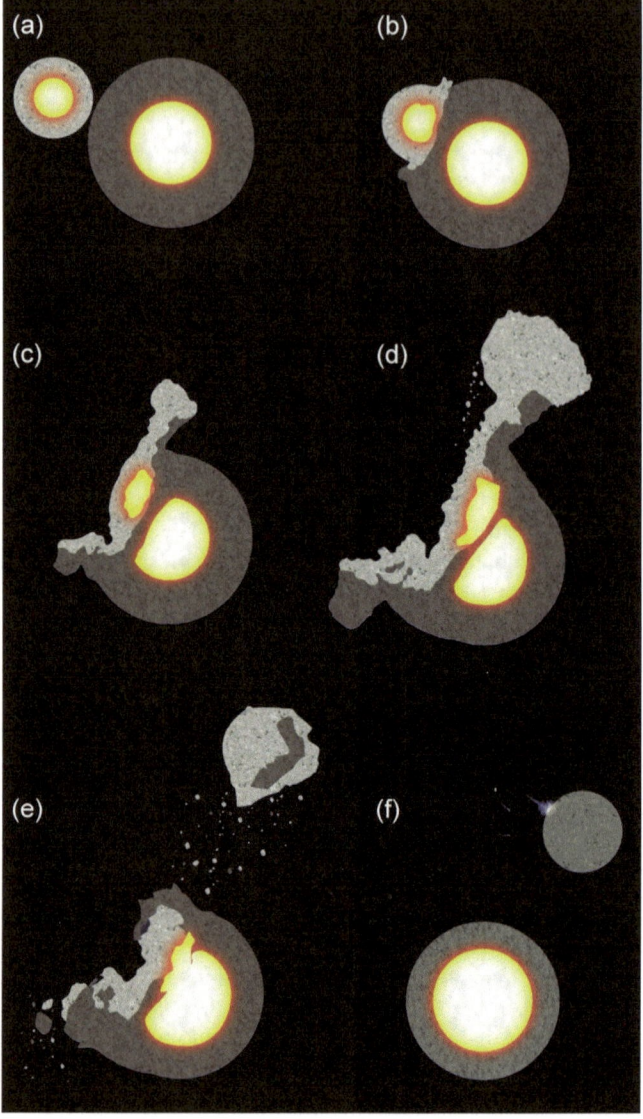

years, the surface of the Earth was a patchwork of lava pools and drifting rafts of recently solidified rock, shrouded in steaming volcanic fumes.

This epoch is referred to as *Hadean* by geologists, after Hades the god of hell, and hell is what the place looked like, with raw lava, boiling sulphur lakes, volcanoes and craters. The atmosphere was an extremely thick brew of carbon dioxide, water vapour and sulphuric acid, laced with dust and soot. Most of the water in the future oceans was still in the air in the form of water vapour, so that the atmosphere as a whole weighed hundreds of times more than today. In fact the pressure at the surface of the planet was similar to that at the bottom of the ocean today, because the pressure depends only on the amount of material overhead, and the weight of water is the same whether it is in liquid or vapour form – vapour just takes vastly more room.

With so much water and carbon dioxide, the greenhouse effect was intense, and the surface temperature was similar to that of present Venus.

Young Earth had become a "steam planet", like some ocean planets that orbit close to their star. The atmosphere, dominated by water, transitioned smoothly from vapour to hot, high-pressure liquid, because the temperature was too high for a well-defined "sea surface" to form (see the phase diagram of water on page 74).

Later meteoritic impacts and lava flows destroyed all trace of this remote epoch in the geological records, and very few pieces of the puzzle remain for geologists to study. On the Moon this epoch can be studied far more clearly, because ancient craters and lava flows have been left untouched.

Earth at 5 – ocean planet

The planet kept cooling, because it still produced much more heat than it received from the young Sun[17]. At some point the temperature dropped low enough at the top of the atmosphere for the first drops of liquid water to condense, forming clouds. Rain began to fall, starting a cycle that is still ongoing more than 4 billion years later.

At first the rain evaporated before touching the ground, because the lower atmosphere was still very warm. But inexorably, pools formed, then lakes, then seas and oceans, until most of the water had condensed out of the air, and the whole planet was covered by a global ocean three kilometres deep on average.

The rain also took out the sulphur from the atmosphere, since sulphur can be dissolved into water as sulphuric acid. The rain in early Earth was acid rain.

Earth had become an "ocean planet". For the first time, it looked Earth-like, with its global blue seas. Its atmosphere of carbon dioxide was still very thick (between two and ten times the present pressure, we are not sure), and rendered opaque by the dust and haze from the volcanic fumes.

Over the next few Eras, landmasses surfaced from the oceans: some volcanoes that erupted from the inside of the Earth were large enough to emerge, like Hawaii, the Canary Islands or the Galapagos today. Some slabs of lighter rocks floated higher on the magma, like a cork on water, to form the first continents.

The rock-carbon cycle, the cycle that integrated the CO_2 in the atmosphere into sediments on continents and under the sea, was very active, and started nibbling away at the carbon dioxide in the atmosphere. Over time, most of the CO_2 was removed from the atmosphere, until nitrogen became the dominant gas.

Somewhere deep in the ocean, around some submarine volcanoes, a strange chemical process got underway. Little pockets of complex carbon molecules started spreading. Basking in the hot water, these primitive cells used the sulphur from the volcanoes to extract chemical energy and prosper.

The water in the ocean was very warm, above 60 degrees Celsius. Early bacteria were thermophiles, which simply means that they liked heat, and could make themselves at home in this global ocean. Over time some of them evolved to master an amazing

---

[17] At the distance of the Earth, the temperature at which heat loss by infrared radiation and heat gain by sunlight balance each other is around zero degrees Celsius.

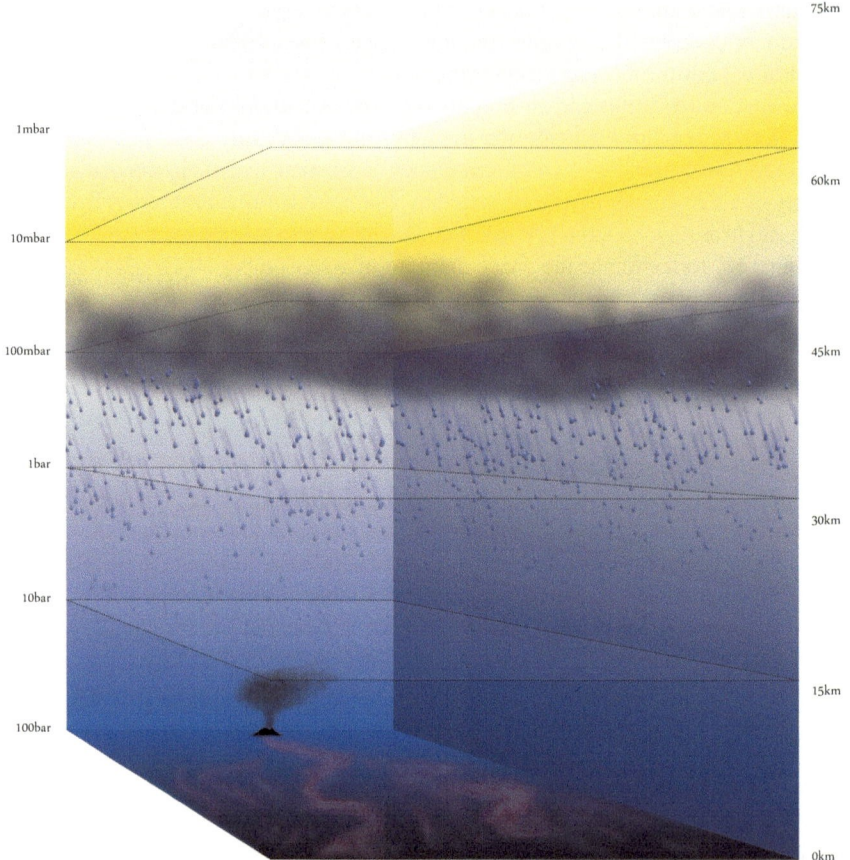

Fig. 8.2 Atmospheric profile of young Earth.

chemical trick: they started using the energy of the Sun instead of the chemical energy of submarine volcanoes and photosynthesis was invented. This allowed them to spread over the whole global ocean, and from then on, life became a factor to be taken into account in the evolution of the Earth's atmosphere.

The engineering of planet Earth by life had started.

Earth at 24 – oxidation and snowball

At 24 Era, 2.2 billion years ago, Earth was a "snowball planet". It was covered by ice and snow from pole to pole, its atmosphere frigid and quiet.

We claimed in Chapter 7 that the climate of Earth was stable in two states, "warm" and "snowball". But what nudged it from one state to the other 2.2 billion years ago?

Apparently, this was the fault of the bacteria that took control of the climate. Photosynthesis breaks water molecules to use their hydrogen atoms as electron donors, and reject the oxygen in the atmosphere; and the oxygen that composes 20 percent of the present atmosphere of the Earth was entirely produced as a poisonous by-product of bacteria and plant photosynthesis. The bacteria which perfected this great chemical feat, the cyanobacteria (called blue-green algae in English, although they are not algae) colonised the oceans, and started rejecting free oxygen on a planetary scale. They were helped in this by the fact that free oxygen was highly poisonous to other forms of life which had not perfected the trick of photosynthesis. Collectively, cyanobacteria produced so much oxygen that from that moment on all the metals at the surface became oxidised. The arrival of photosynthetic cyanobacteria is marked in the geological records by the disappearance of metallic iron formations in the rocks and the global dominance of rust. But the oxygen content in the atmosphere did not rise at the beginning of the era of cyanobacteria. At first, there were enough minerals and metals to oxidise on the ground for all the oxygen produced to be immediately returned to the rocks.

Only when most rocks were finally oxidized did the content of free oxygen in the atmosphere start to rise, to about 0.1 percent. Oxygen-using bacteria started appearing, as well as more complex, but still single-celled life forms.

As soon as oxygen became sufficiently abundant, it cleared the methane out of the atmosphere. Methane is a very efficient greenhouse gas, and oxygen is not, so this caused a global drop in temperature – enough to nudge the planet into a positive feedback loop (increasing white snow and ice cover causing more sunlight to be sent back into space, which cooled the planet and further increased the snow cover) that lead to the snowball-Earth episode.

The Earth was now a frigidly cold planet, almost entirely covered by a giant ice sheet. Only one patch of ocean remained clear of ice near the Equator. Life struggled to survive near the ice margin and in the open oceans. Close to the poles the temperature was forbiddingly low throughout the year.

Earth at 35 – the boring billion

At 35 Eras, we catch the Earth in what geologists call "the boring billion", because nothing much happened to the rocks in a billion years.

The planet had recovered long ago from the snowball episodes. It was saved by the volcanoes. The ice covering the land and oceans blocked the rock-carbon cycle: the carbon dioxide in the air could no longer be integrated into rocks. But the other side of the cycle – the injection of CO_2 into the air by volcanoes – kept on going; volcanoes had no problem piercing through ice, as they do in Iceland today. Over time, the carbon dioxide concentration in the atmosphere slowly increased, until the greenhouse effect was high enough to melt the ice again.

The oxygen content in the atmosphere was stable at around 0.1 percent. This was not enough to produce an ozone layer. Therefore the stratospheric lid that kept the clouds below ten miles in the present Earth did not operate.

Convection must have extended much higher into the atmosphere. This would have been called "the Great Age of Clouds" by any visitor, since the high temperatures, large oceans and lack of stratospheric lid would have produced the most magnificent cloud formations and the most awesome storms.

During this stage, Earth was a living planet, recognisably Earth-like with oceans and continents, and vast colonies of blue-green algae. Life had been present for at least three billion years, yet there was no signature in the atmosphere that would make it detectable for an alien astronomer. This is a sobering thought in the search for inhabited exo-Earths: life on our own Earth has been chemically discreet for 90 percent of its tenure.

The convective motions in the Earth's mantle pushed the continents around, and every few hundred million years, they gathered together in a single super-continent, then they broke up again into half a dozen pieces or so. The last super-continent was Pangea, where the dinosaurs roamed. At 35 Eras, landmasses gathered into one such super-continent, Rodinia.

Continental drift provides the second half of the rock-carbon cycle, the one that allowed the Earth to recover from the snowball episodes: oceanic plates are buried into the Earth's mantle by tectonic motions. The carbonates that they contained are burnt and turned back into CO_2, which escapes from the volcanoes back into the atmosphere. By modifying the amount of volcanic activity, as well as the surface of land available for the formation of carbonates, the shape and displacement of continents modifies both sides of the rock-carbon cycle, and therefore influences the amount of CO_2 in the atmosphere and the evolution of the climate (in addition to the more obvious direct effect on weather and sea currents, see Chapter 7).

Earth at 40 – animal planet
Finally, around 6 Eras ago (at 40 Eras), the level of oxygen had risen sharply again, until it reached 20 percent of the total. We have become used to such air, but as we realised in Chapter 1 this is really a massively explosive atmosphere.

A new type of large organism called plants evolved to take advantage of such an oxygen-rich atmosphere. During the day they used sunlight to produce energy and exude oxygen (like cyanobacteria), but during the night they breathed oxygen and burnt it as a further source of energy. Another class of organism, that we call animals, lived entirely on the free oxygen. They survived by feeding on the plants and using some of the oxygen rejected by photosynthesis to power their high-activity lifestyles.

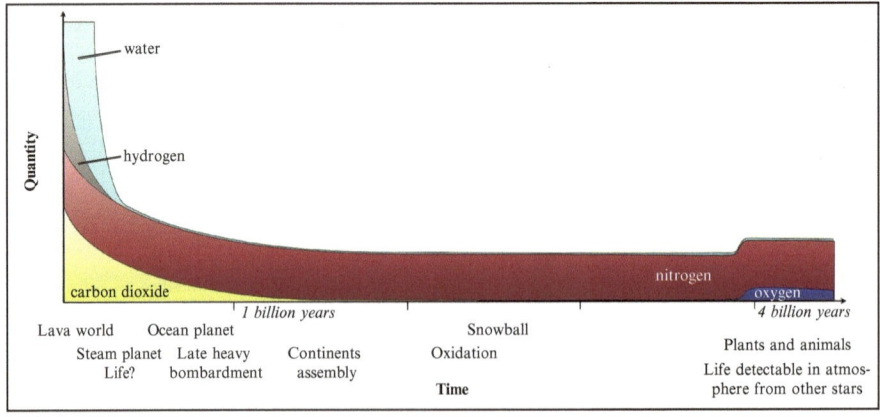

Fig. 8.3 Evolution of the composition of Earth's atmosphere over time.

Fig. 8.4 Earth at 40.

Incidentally, it is interesting to realise that animals had no impact at all on the composition of the atmosphere[18].

About 3 Eras ago (Earth is now 43) we arrive at the kind of world that most animals have known. The continents have finally been colonised by life, giant vertebrates roam the land (the dinosaurs are still in the future). Because there is still more CO_2 in the air, the climate is warmer than today. Both poles remain ice-free.

Earth at 46 – Ice ages

We now reach Earth at 46 – today. We catch it in the middle of its Ice Ages. It may not look that way to us, but from a global perspective the Earth is now in a short, not particularly warm interval between two Ice Ages. For a few million years, it has been oscillating between slightly colder episodes, when the ice sheets extended over most of Europe and North America, and warmer episodes, with ice mainly over the Arctic and Antarctica. Recall that for most of Earth's history the poles had been ice-free, and, snowball episodes excepted, the Earth has never been as cold as in the present geological era. This is due to the "recent" collision of India into Asia, which started about 50 million years ago (1/2 Era) causing the rise of the Tibetan plateau, which is the dominant tectonic event of our Era. It turns out that the high plateau and the monsoon phenomenon that it triggered are very efficient at increasing the rock-forming part of the rock-carbon cycle. The drop in CO_2, together with other tectonic changes such as the opening of the Drake passage, have lowered the global temperature of the planet enough to keep the poles permanently frozen.

Soiling the pool

However, over the last hundred years, human beings have been pumping CO_2 back into the atmosphere at an incredibly fast rate. So fast that the concentration of CO_2 has doubled in only 16 years, or 0.000016 Era, a geological eye blink. The CO_2 concentration has increased from 150 grammes per tonne in 1850 to 300 grammes per tonne in 2010. This is as it was 60 million years ago, at the dawn of the age of mammals. As it

---

[18] With possible exceptions. The first, of course, is when one animal species starts burning fossil fuel and putting carbon buried underground back in the atmosphere. Another one is weirder: recent research suggests that a large part of the marine carbonate rock deposits, such as the White Cliffs of Dover, may result from carbonates precipitated in the guts of fish. Therefore ocean fish may play a part in the carbon-rock cycle and help remove the CO_2 out of the atmosphere.

takes more than a hundred years to heat the ocean, the climate hasn't returned to the temperature it was in the Eocene, not just yet.

Fig. 8.5 Soiling our pool.

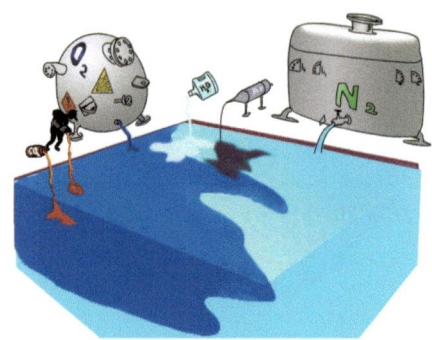

How can such a small amount of gas, less than one part per thousand, have such an impact? Let us think in terms of our "pool" image (Fig. 8.5). A public swimming pool contains about one million litres of water. 100 grammes per tonne is therefore 100 litres. Imagine leaking 100 litres of mildly harmful substance into the swimming pool. It isn't like one swimmer discretely peeing in the pool, but rather like all swimmers doing it for the whole day.

In global terms, human-induced climate change will just be a blip like the Eocene volcanic event. When the injection of CO_2 finally abates, as it has to because the reserve of fossil fuels is finite, the "volcanic" side of the rock-carbon cycle will go back to normal, and the excess CO_2 will be turned back into rocks (not coal and petrol, this time, but calcium carbonates).

Venus is sometimes taken as an example of how global warming could turn very ugly, sending the Earth's climate in a runaway feedback state that would lead to the evaporation of the oceans and the loss of water to space, but we are probably too far from the Sun for that.

Earth's future

Stars like the Sun live for about ten billion years, before turning into red giants. Over their lifetime, they become slightly brighter: the Sun is about 30 percent brighter now than it was for the young Earth, and will be another 30 percent brighter in the next five billion years.

About ten Eras into the future, the oceans will have become so warm that complex life will be cooked. 30 Eras into the future, the oceans will start evaporating, sending the planet onto the "runaway greenhouse" slope leading to a Venus-type atmosphere. The planet will thus retrace some of its steps, a hot ocean, then steam planet, then with a CO_2 atmosphere, until the Sun turns into a red giant star.

Since the death of the Sun probably implies the destruction of the Earth, the story may end here. But for most planets in the Solar System, the demise of the Sun will be only one event among others. The core of the Sun will turn into a white dwarf, and

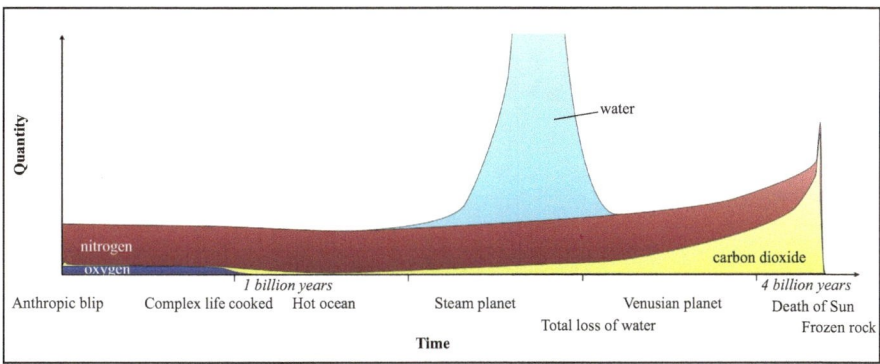

Fig. 8.6 Possible evolution of the composition of Earth's atmosphere in the future.

Mars for instance will keep orbiting it indefinitely. Its atmosphere will freeze out, as will that of Titan.

One interesting twist in the story of the future of the Solar System is that, during the red giant stage, the icy satellites of Jupiter and Saturn will become warm enough for ice to melt, and for water oceans to form at the surface. The red giant stage of the Sun may be short compared to its total lifetime, but it will still last several hundred million years. We may speculate that this is long enough to start life on these planets – it is certainly long enough to see it prosper, if it has already started in the under-ice oceans. The red giant stage of Sun-like stars may therefore offer short abodes for life in the outer edges of planetary systems.

If Earth has had so many incarnations in its long life, what about its close companions? It is harder to reconstruct the ancient climates of Venus and Mars than that of Earth because we have far less data. But as far as we can see, it is clear that both planets were also very different in the past compared to what they are now.

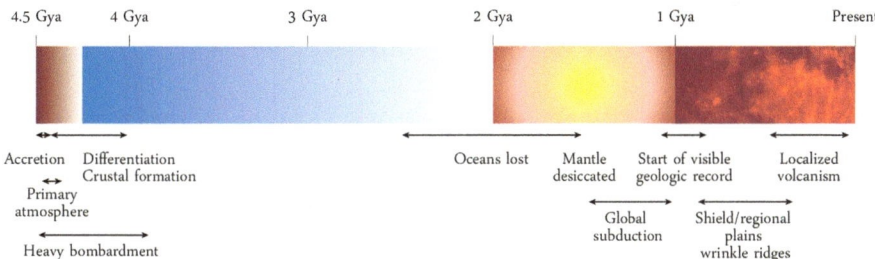

Fig. 8.7 A possible reconstruction of the evolution of Venus. Some specialists think that Venus could have been covered by water oceans for the majority of its life, losing them relatively late. Image credit: David Grinspoon

Epilogue

Now that we have crossed many eons of time and light-years of space, wandered in all corners of the periodic table, and experienced extremes of pressure and temperature, let us take another look at the Earth's atmosphere, one serene spring day in a temperate region.

What do we see? A grand procession of clouds in a blue sky, some looking like animals, some making faces?

With our new "Alien Skies" glasses on, the landscape may look different.

We see a nitrogen atmosphere, with a huge fraction of reactive oxygen, and a layer of ozone overhead that blocks the incoming ultra-violet rays and prevents the clouds from rising into the stratosphere. Droplets of water form clouds, because vast atmospheric swirls, formed when thermal currents try and fail to move from Equator to Pole, mix warm air with colder air. A thin haze of aerosols – small particles of dust, sand, soot or ash: gives a reddish tinge to distant views. On the surface, an extraordinary green colour signals the ubiquity of photosynthetic life on this planet, explaining the continued presence of the free oxygen.

Fig. 8.8

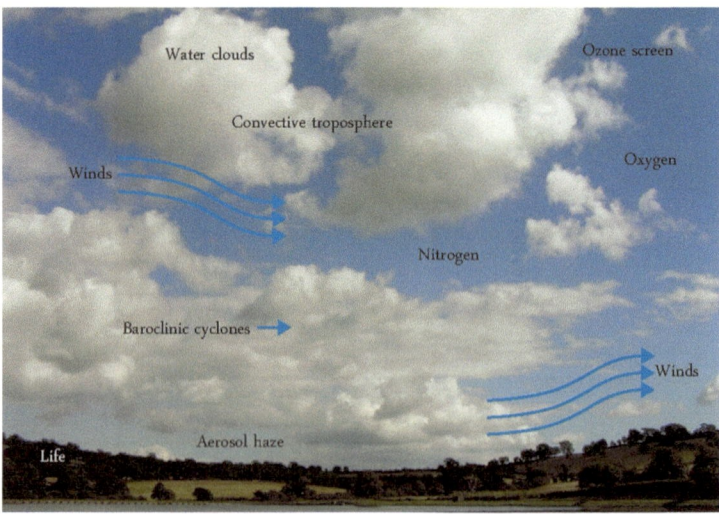

With our "Alien Skies" glasses, we realise how this atmosphere is one of so many possibilities: blue skies on some planets spell carbon dioxide; clouds can be made of methane; storms can be due to winds blowing from day side to night side; on a tidally locked planet the stratosphere can be heated by titanium oxide; the Sun can be green because of sodium atoms; the cliffs can be made of silicon carbide; the haze can be due to ruby dust. As for life on other planets, we still don't have much of a clue ...

Fig. 8.9

What are the limits? How alien can alien atmospheres be? Is there anybody to watch them? These are early days in the great voyage of discovery of planetary atmospheres.

Fig. 8.10

Notes on sources and complementary information

exoplanet names (page 2)
The International Astronomical Union has decided not to attribute individual names to exoplanets. That seems prudent; I suspect that the day naming was allowed, the sky would instantly fill with women's names and we would have to give scientific papers such titles as "Samantha much hotter than previously thought". This impression was confirmed when in 2010 an astronomer named his most high-profile planet discovery (subsequently shown to be spurious) with his partner's name. Catalogue names and numbers may not be ideal, but they become familiar quickly enough, and I don't think the name *1644* has been a handicap for the French lager of that name.

Nevertheless, I found it impossible to write this book with the official planet names. The two most well-studied exoplanets are called HD 189733 b and HD 209458 b, and a paragraph filled with these two jumbles of digits quickly becomes confusing. That is why I used names – a sin that may bring the ire of the community on my head. For HD 209458 b, I was helped by Alfred Vidal-Madjar at the Institut d'Astrophysique de Paris, who in defiance of the IAU calls the planet Osiris in his papers about it. Alfred has been the first to identify atomic lines in the cometary tail trailing the planet. I made up the other names using the same Egyptian deity theme. The only objective is to make the text more readable, and I apologise to the reader if that means it will be more difficult to spot these planets in the news. The following table gives the correspondence.

Osiris	HD 209458 b
Isis	HD 189733 b
Nephthys	HD 80606 b
Anubis	GJ 1214 b

the first exoplanet (page 2)
The first known exoplanet is usually said to be *51 Peg b*, discovered by the team of Michel Mayor at the Haute-Provence Observatory in France in 1995. "The first exoplanet around a normal star" is the full statement, because three years earlier planets were found orbiting a neutron star by Aleksander Wolszczan and his team using the Arecibo radio telescope in Puerto Rico. A neutron star is the dead core of an exploded supernova, and these bodies would be very different kinds of planets. We may also have to add "first *normal* planet" to the 51 Peg statement, since in 1989 a team lead by David Latham at Harvard found an invisible companion to the star GJ 229 that they then labelled as a brown dwarf, but that we would today classify as a planet. It was none other than team member Michel Mayor who argued at the time that the companion of GJ229 could not be a planet, because its orbit was too eccentric. As we see in Chapter 8 of this book, it turned out that many planets have eccentric orbits, contrary to what was thought at the time.

There are several books on the early planet hunters, including *Planet Quest* by Ken Croswell.

false alarms (page 2)

The detection of the first planet outside the Solar System was announced several times before 1995. Famous examples include the case of Barnard's Star. Fluctuations in the position of this nearby star were interpreted by Dutch Astronomer Piet van de Kamp as the signature of an unseen Jupiter-like planet. The community remained unconvinced, but it took a very long time to dismiss the hypothesis entirely. The double star system *61 Cygni* was also at some point thought to harbour a massive planet. Spurious detections remain common today. It is a fact of science that seeing signals where there is only noise is easier than accumulating the much larger amount of high-quality data required to prove the absence of a signal.

seemed to cast doubt on the whole issue (page 2)

This is a paper published in 1997 by Canadian astronomer David Gray in *Nature*, showing that the apparent velocity change in 51 Pegasi – interpreted as a planet signature – was associated with changes in the shape of the spectrum itself. This is not supposed to happen if the signal is due to a planet, but will happen when due to a pulsation of the star. The objection carried special weight because Gray was a top specialist of the spectroscopic technique, author of a reference textbook (*The Observation and Analysis of Stellar Photospheres*). The data used by Gray, though, was very noisy, and the analysis a bit shaky. As more hot Jupiters were discovered by the team of Geoffrey Marcy in California, Gray's "exotic stellar pulsation" lost support. The hypothesis was never entirely disproven, it just gradually vanished over time the way incorrect scientific results tend to do.

transit of Osiris (page 2)

The transit of HD 209458b (Osiris in this book) across its star was first measured by David Charbonneau and Timothy Brown with a four-inch telescope (this is no misprint, four inches only) at Harvard, or by Gregory Henry from Arizona. The story is entertaining, at least in the fictionalised account which was written in French as a novel by Elisa Brune, in *Les Jupiters chauds*.

The planet itself was first detected using the Doppler-wobble method by Tsevi Mazeh from Tel Aviv University, working with planet-search specialists Dave Latham and Michel Mayor. The one percent dip in luminosity is relatively easy to detect even with a small telescope if you know when to observe (when the planet is aligned with the star), and that is what the two teams of observers did. Charbonneau and Brown then went on to observe the transit with the Space Telescope, thus obtaining the first measurement of an exoplanet atmosphere with the detection of the sodium line in the atmosphere of Osiris. David Charbonneau is now professor at Harvard University and Tim Brown leads a network of privately funded telescopes based in Santa Barbara, California.

MOLA (page 3)
The MOLA altimeter is an instrument aboard the NASA Mars Global Surveyer mission. Between 1996 and 2001, it has measured the altitude of the Martian terrains with a vertical resolution of 30 metres, producing a detailed relief map of the whole planet. The maps are available online on the MOLA website. There are many fantastic books on the mapping of Mars, including *Mapping Mars* by Oliver Morton.

talk by Raymond Pierrehumbert in Oxford (page 4)
This was the *Halley lecture*, an annual event at the University of Oxford. Raymond Pierrehumbert is the author of a wonderful textbook on planetary atmospheres, *Principles of Planetary Climate* (for scientists), and gives very inspiring conferences.

Exoclimes conferences (page 4)
The first "Exoclimes" conference was organised by the author, together with two fellow astrophysicists, Suzanne Aigrain (now at Oxford University) and Isabelle Baraffe, in Exeter in September 2010. Webcasts of the review talks and the slides of all talks are available online at *www.exoclimes.org*. The organisation of the second Exoclimes conference was lead by Nick Cowan at Northwestern University (Chicago). It took place in January 2012 in Aspen, Colorado. The slides from the talks are available on the same website. The third Exoclimes conference is scheduled for February 2014 in Davos, Switzerland. On the associated *www.exoclimes.com* website, researchers from Exeter and Oxford provide short summaries of the latest research developments in the field of exoplanetary atmospheres.

Bertrand Piccard (page 10)
Bertrand Piccard is the latest in a distinguished dynasty of adventurous scientists. His grandfather was Auguste Piccard, who inspired the Professor Calculus character in the Tintin stories, and his father was Jacques Piccard, who built the Bathyscaphe submersible to reach the deepest point in the oceans. We can expect one of his three daughters to be the first human to explore an alien atmosphere.

Global conveyor belt (page 20)
The global circulation of the ocean currents is also called *thermohaline circulation*, from the Greek thermo for heat and haline for salt. Although the difference of heat between the Equator and the poles is the main driver, the injection of salt-free water from the melting icecap in the Arctic plays a key role. Salty water is heavier than freshwater, so the warm but salty water from the Equator sinks below the cold water from the poles, closing the loop of the current around Iceland. Thus the pattern of global ocean currents depends not only on the Equator-to-pole heat transfer, the shape of the continents and the Coriolis effect, but also on the extent and position of the icecaps at the poles.

Fig. 2.8 (page 33)
This reconstruction is due to Don Mitchell, from his blog/website called *Mental Landscape.*

Carl Sagan (page 49)
The "nuclear winter" simulations were published in *Science* by five US scientists (Richard Turco, Owen Toon, Thomas Ackerman, James Pollack and Carl Sagan) in 1983.

Fig. 3.14 (page 51)
This figure comes from the work of François Forget and his group, published in *Science* in 2006. François Forget, from the Laboratoire de Météorologie Dynamique in Paris, is one of the pioneers of using climate models for planets other than Earth.

uniquely among moons, Titan possesses an atmosphere (page 55)
Some other bodies in the Solar System have very tiny amounts of gas around them that can be called an atmosphere in some contexts. Pluto and Triton for instance have a few millionths of a bar of gas above their surface. This is sufficient to show winds and even clouds. But in the context of exoplanets such tiny amounts of gas do not have a sufficient effect on the planet as a whole to qualify as real atmospheres in the sense of this book.

the GAIA view of the Earth as a system (page 65)
The "GAIA" concept is a stimulating and controversial way to consider the atmosphere, ocean and biosphere of the Earth as a whole, introduced by British scientist James Lovelock. The basic idea is that the system has become complex enough to react like a giant living organism and regulate its own environment to ensure its continued existence. Lovelock elaborated on his concept for the general public in a series of books: *Gaia, the ages of Gaia, the revenge of Gaia* and *the vanishing face of Gaia*, with titles reflecting increasing alarm at mankind's assault on the ecosystem.

The *Daisy World* computer simulation was introduced in 1983 by Lovelock and his then student Andrew Watson (now professor at the University of East Anglia).

in the Iliad (page 68)
The passage is in Book 8:
You will see how much mightier I am than you immortals. Go on: attempt it, and see. If you tied a chain of gold to the sky, and all of you, gods and goddesses, took hold, you could not drag Zeus the High Counsellor to Earth with all your efforts. But if I determined to pull with a will, I could haul up land and sea then loop the chain round a peak of Olympus, and leave them dangling in space. By that much am I greater than gods and men.

Fig. 5.13 (page 79)
This figure comes from the work of Peter Read at Oxford University, a planetary scientist who studies the climates of Venus and Mars, and in particular the effect of rotation on the flow in planetary atmospheres.

the NASA Kepler planet search mission (page 83)
Kepler is a space mission launched by NASA in 2009 to conduct the census of exoplanets down to Earth-sized planets using the transit method. The Kepler probe is orbiting the Sun some distance behind the Earth (it is therefore not orbiting the Earth directly, but slowly drifting away from us). Its array of cameras is taking an image every fifteen minutes of the same area of the sky, a field of around 150,000 stars in the direction of the constellation Cygnus. The Kepler mission has detected thousands of transiting planets, and has established the frequency of planets around stars like the Sun, depending on size and orbital distance. More details can be found on the mission's website.

Hot Jupiters (page 84)
There is no official definition of the term *hot Jupiter*. The term usually designates gas giant planets with orbital distances below 0.1 AU, i.e. nearer to their star than a tenth of the Sun-Earth distance. The temperature of a planet at this distance will depend on how hot the host star is. Some researchers attempted to introduce other names for such planets, including "Roasters" and "Vulcans", but only the simple and descriptive term "hot Jupiter" has gained general use, with its remarkably unpoetic extension "very hot Jupiter" for extreme cases with periods shorter than around two days.

shockwaves (page 85)
The question of shocks in the atmosphere of hot Jupiters is still unresolved. I take my cues from the work of Kevin Heng at Bern University and Adam Showman at the University of Arizona, but not all specialists would agree, and there are yet no direct observations of the effect of shocks in hot Jupiter atmospheres.

cometary tail (page 86)
The notion of an extended tail of escaping gas around Osiris comes from the work of Alfred Vidal-Madjar and his team at the Institut d'Astrophysique de Paris. In 2004 they used the Hubble Space Telescope to observe the lines of hydrogen, oxygen and carbon in the ultraviolet, and found that these lines were enhanced during the transit, which could be interpreted as the signature of a large amount of gas escaping from the planet. At first, this was thought to imply that hot Jupiters were doomed to evaporate over the life time of stars. With more hot Jupiters known, better measurements and more detailed models of atmospheric escape, it is now thought that the amount of gas escaping from planets such as Osiris is not sufficient to affect the planet as a whole.

Alkali metals (page 88)
There are many great books about the periodic table, including *Periodic Tales* by Hugh Aldersey-Williams and the classic *Uncle Tungsten* by Oliver Sacks. Also check out the fantastic *periodic table of videos* produced by the University of Nottingham's chemistry department, for vivid first-hand experiences with the elements.

sunsets (page 89)
I was part of the team that discovered the planet Isis in 2005, and when the different types of follow-up observations were divided up, I took charge of the measurement of the transit with the Hubble Space Telescope, to measure the size of the planet precisely. As a by-product of these observations, we could check whether the size of the planet varied with wavelength because of the signature of the atmosphere. Models predicted that the spectroscopic features of the atmosphere would be too small to be detectable, but I carried out the analysis anyway – it is a general rule for scientific observation never to trust the models prediction entirely.

What I found instead, after much work to account for the variations of the star itself due to its spots, was the clear and unexpected signature of scattering by a haze of small particles. Alain Lecavelier at Paris Observatory then calculated that the most likely constituent of this haze was transparent grains or drops of enstatite, a type of magnesium silicate.

Later on, with David Sing at the University of Exeter and other specialists of exoplanet atmospheres from the United States (Ronald Gilliland at NASA, Heather Knutson at Caltech and others), we collected more observations with the Space Telescope that confirmed this picture, and gathered the information to reconstruct the visible colours of a sunset seen through the atmosphere of the planet. Most recently, with Tom Evans at the University of Oxford, we detected the direct reflection of the planet in visible light (as opposed to the transmission spectrum obtained during the transit). These observations suggest that the planet has a deep blue colour when seen directly.

More details can be found on www.exoclimes.com, or with search terms like "sunset on HD 189733b" or "exoplanet color" on the internet.

a telescope (page 98)
The planet mentioned here was identified by the OGLE transit search. OGLE is an observation programme set up by the late Bogdan Paczynski, professor of astrophysics at Princeton University, to detect dark matter in our Galaxy. The method is to monitor a very large number of stars in search of the very rare events in which lumps of dark matter would focus the light of a star and make it shine brighter for a few days, an effect expected from Einstein's theory of general relativity, called "gravitational lensing". As a by-product of this programme, the OGLE telescope at Las Campanas Observatory in Chile was also used to look for transiting planets. It detected the first exoplanet identified by its transit, OGLE-TR-56b, then five other planets, before being overtaken by transit surveys specifically designed to look for planets, such as *HATNet* and *superWASP*.

brightness map (page 99)
The brightness map of Isis was obtained by a group led by David Charbonneau and his PhD student Heather Knutson at Harvard University. In 2007 they used the infrared camera aboard the Spitzer space telescope to follow the phases of Isis over a full orbit, and reconstructed the distribution of brightness on the surface of the planet from the brightness variation along the phase curve.

HR 8799 (page 101)
The system HR 8799 was discovered in 2008 by a team led by Canadian astrophysicist Christian Marois using the Keck and Gemini giant telescopes in Hawaii. They obtained very precise images of the star and its immediate neighbourhood by a technique known as "adaptive optics", whereby ultra-rapid motion of mirrors in the telescope compensate for the turbulence of the atmosphere to produce images almost as sharp as those taken from space. They then needed to post-process the images with computers to remove the dominant light from the star and detect the tiny dots of light from the planets. They could prove they were planets because the dots moved around the star with time in the manner predicted by Kepler's laws.

upheaval in the Solar System (page 106)
This model of the orbital evolution of the Solar System is presented in a series of papers by A. Morbidelli, K. Tsiganis, R. Gomes and H. F. Levison in *Nature*. An animation in this scenario is available online. Not all specialists consider that the scenario has been proven yet.

we seem to be safe (page 106)
The calculations about the orbital evolution of the Solar System are by the group of Jacques Laskar at the Paris Observatory. These calculations take enormous amounts of computer time, and as far as I know the Paris group does not have direct competitors, which is never a good thing in science. So these conclusions may still change, as they have done in the past. One can argue that predictions about the future of the Solar System in billions of years time do not belong to the realm of science anyway, since they will remain untestable.

Anubis (page 108)
GJ 1214 was observed by many teams using telescopes in space and on the ground to try to measure its atmosphere, but the task is difficult. Most results converge on the conclusion that the atmosphere is not detected at any wavelength, which tends to indicate that it is either heavy or clouded. But negative proofs such as this one are never as good as positive evidence, and we probably will have to wait for the next generation of telescopes and instruments, around the year 2020, to draw definite conclusions. See *Exoplanet Atmospheres* by MIT Professor Sara Seager for a technical book on this topic.

Steppenwolf (page 113)

Steppenwolf is a 1927 German novel by Herman Hesse, where the main character is a loner who feels cut off from everybody else. It was chosen to designate free-floating planets with buried oceans by Dorian Abbot at the University of Chicago. Planets can be ejected from systems by gravitational interaction with other planets or with another star. While no such case has been established with certainty, it is likely to be quite common if our current understanding of planet formation as a very violent process is correct.

tidally synchronised Earth (page 115)

To produce Fig. 7.13 I have used the output of a circulation model of a tidally synchronised, Earth-like planet by Kevin Heng (astrophysicist at the University of Bern), and superimposed it on the map of Earth in the absence of rotation according to Witold Fraczek at ESRI (a technology company based in New York). For the longitude of the centre of the day side I have used the heaviest region according to measurements of the local gravity anomalies on the globe. I have estimated the influence of humidity by eye (i.e. more humid implies less extreme temperatures).

pool planet (page 116)

The "pool planet" case was calculated by Raymond Pierrehumbert from the University of Chicago (who now calls it *eyeball planet*). It corresponds to a tidally locked Earth that becomes cool enough for a global glaciation to take hold, except near the centre of the day side. If ocean planets are common, such cases could be frequent as well, around stars cooler than the Sun (which are the most common type of stars). In such cases, a planet would have to be closer to remain warm enough to sustain liquid water, close enough for tidal locking to operate.

Nephthys (page 117)

HD 80606b was identified in 2001 by Dominique Naef, a student of Michel Mayor, at the Haute-Provence Observatory in France. Its extremely elongated orbit made the detection challenging: it barely tugs on its star for most of its 111-day orbit, then furiously pulls it during a few days at closest approach, before resuming its stately journey. Checking whether the planet was transiting was another challenge, because of the long period: just one short event every 111 days. On 13th February 2009 I was at Haute-Provence Observatory with one of the team who caught the event. A memorable night.

in the sky of Nephthys (page 118)

I have used a plot of the orbit of HD 80606b by US astrophysicist Greg Laughlin to calculate the figure. For some years Greg kept a very entertaining, extremely reliable and beautifully illustrated blog on exoplanets (oklo.org). It is still worth a visit.

diamond planets (page 119)
The 'diamond exoplanet' idea was introduced by NASA astrophysicist Mark Kuchner, who explored the consequences of a higher abundance of carbon than oxygen on the internal structure of a terrestrial planet. It was taken up by the press with the usual mixture of enthusiasm, simplification and short attention span, and the idea of "diamond planets" briefly circled the world, with a few shiny illustrations.

exo plate tectonics (page 120)
Work on the behaviour of plate tectonics on planets larger than Earth was initiated by Dimitar Sasselov at Harvard University. Geophysicists are quick to point out that on Earth the behaviour of tectonic plates can depend very sharply on details, such as the presence of tiny amounts of water in the mantle to "lubricate" the motion of plates, and that it is therefore probably vain to try to derive some general behaviour for exoplanets. Exoplanet researchers would argue that pushing the models out of their comfort zone is often a good way to test our understanding, even in the absence of data.

life (page 121)
The two mission concepts from the 1990s to measure signs of life in the atmosphere of an exoplanet are NASA's "Terrestrial Planet Finder" and ESA's "Darwin". There is, of course, plenty of literature on the topic of the search for exo-life, including the wonderful *Lonely Planets* by David Grinspoon.

Epoxi (page 124)
Epoxi is a reincarnation of NASA's *Deep Impact* mission. If you missed it, *Deep Impact* was a movie in which Mel Gibson uses a big drill to fight a comet and save the world. The *Deep Impact* probe travelled to a comet called *Tempel 1* and on July 4 (sic) 2005 dumped a fridge-sized load onto the comet to observe the explosion and study the composition and porosity of the surface. Its mission accomplished, the probe was then redirected towards another comet, which it reached in 2010. NASA scientist Drake Deming and others proposed to use the probe's camera during the trip between the two comets to study exoplanets. This included observing the Earth as an exoplanet – the probe was by that time so far from our planet that it only saw Earth as a point of light in the sky. The brightness of the Earth was measured over two days in two colours, and Nick Cowan at Northwestern University led a study of this data, showing that it enabled a reconstruction of the broad distribution of landmasses and oceans on Earth, in a paper enigmatically entitled *Alien Maps of an Ocean-Bearing World*.

Earth as a snowball planet (page 130)
The extent and duration of the "snowball episodes" for Earth are hotly debated in the scientific community. While most researchers would agree that the geological records indicate periods of glaciation much more severe than the recent Ice Ages, the idea of an entirely frozen planet is not accepted by all. A large swathe of the oceans near the

Equator could have remained ice-free. The number of cold episodes is also debated.

Hydrogen atmosphere (page 127)

It is very difficult to reconstruct what the earliest states of the Earth's atmosphere might have been. The earliest rocks dated by geologists are around four billion years old, a full 500 million years after the formation of the planet. They have not conserved inclusions of bubbles of air (the way Antarctica ice cores allow us to trace the composition of the air in the more recent geological past), so that the composition must be inferred from very indirect clues. Hydrogen in the early atmosphere has left no trace, and is only inferred from our present understanding of the formation of terrestrial planets. Every aspect of the composition of the early atmosphere is highly speculative, including the total amount of carbon dioxide. As usual for this book I have used what I take to be the "reasonable current best-guess".

Earth at 40 (page 132)

Nice drawings of the reconstruction of the Earth's map in the deep past can be found at the Paleomap Project of Christophere Scotese. These reconstructions are based on a very large number of geological studies, using a wide variety of techniques ranging from recognising common rocks in presently widely separated regions, and tracing the distribution of fossil species, to the measurements of the magnetisation of rocks that keep a memory of their ancient position relative to the magnetic field. The usual disclaimers apply: many details are speculative, and the more we go back in time, the hazier our knowledge becomes. Nevertheless, geologists have now become extremely good at that peculiar type of spherical puzzle-solving.

soiling the pool (page 133)

The concentration of CO_2 in the atmosphere has been monitored by the Mauna Loa Observatory in Hawaii since 1958. It has risen from 315 grams per tonne in 1958 to 390 grams per tonne in 2012 and keeps rising at an increasing rate. This number was around 200 during the ice ages, about 760 at the dawn of the age of mammals, and 1000 to 3000 during the age of the dinosaurs.

Earth's future (page 134)

Conclusions about the future of the Earth are very speculative, and venture close to the limits of science because there is no way to verify them with experiments. I have based my remarks on the talks presented on this topic during the Exoclimes conferences, and at the NASA conference on planetary climates in 2012 in Boulder, Colorado. Opinions have been shifting, particularly on the issue of how much time will elapse before our planet becomes uninhabitable because of the increased solar luminosity.

Venus timeline (page 134)

The Venus timeline comes from the talk by David Grinspoon at the Exoclimes 2010 conference. Dr Grinspoon is curator at the Denver Museum of Natural History and a specialist on the atmosphere of Venus. His book *Venus Revealed* on the topic can be highly recommended.

Moon-forming impact (page 128)
Although the impact scenario for the formation of the Moon is considered likely, other scenarios are still posited by the scientific community. The main argument for the impact scenario is that the composition of Moon rock is similar to that of the Earth's crust, and that the Moon's total weight indicates that it lacks a heavy metallic core, as expected if it formed from ejected bits of the Earth's crust.

Index